Hare Brain Tortoise Mind

Hare Brain Tortoise Mind

How Intelligence Increases When You Think Less

Guy Claxton

THE ECCO PRESS
An Imprint of HarperPerennial
A Division of HarperCollins*Publishers*

For Jo

First published in Great Britain in 1997 by Fourth Estate Limited.
A hardcover edition of this book was published in 1999 by The Ecco Press.

HarperCollins books may be purchased for educational, business, or sales promotional use. For information please write: Special Markets Department, HarperCollins Publishers Inc., 10 East 53rd Street, New York, NY 10022.

First HarperPerennial edition published 2000

Text design by Richard Kelly
Text figures drawn by Carole Vincer

Library of Congress Cataloging-in-Publication Data has been applied for.

ISBN 0-06-095541-4 (pbk.)

06 07 08 09 RRD 10 9 8 7

Everything is gestation and bringing forth. To let each impression and each germ of a feeling come to completion wholly in itself, in the dark, in the inexpressible, the unconscious, beyond the reach of one's own intelligence, and await with deep humility and patience the birth-hour of a new clarity: that alone is living the artist's life. Being an artist means not reckoning and counting, but ripening like the tree which does not force its sap, and stands confident in the storms of spring without the fear that after them may come no summer. It does come. But it comes only to the patient, who are there as though eternity lay before them, so unconcernedly still and wide.

Rainer Maria Rilke

Contents

Acknowledgements

There are many people who have supported, encouraged and guided me throughout the long gestation of this book, and to whom thanks are due. They include Stephen Batchelor (for duck dinners), Mark Brown (for his enthusiasm), Merophie Carr, Polly Carr, Isabelle Gall, Rod Jenkinson (who wanted it to be *The Sin of Certainty*), Kikan Massara, Helen and Colin Moore (for their room), and my mother, Ruby Claxton (for her love and for not interrupting). Among those who offered me scholarly advice and generously shared their time and knowledge were Peter Abbs, Maurice Ash, Brian Bates, Susan Blackmore, Alan Bleakley, Laurinda Brown, Fritjof Capra, Martin Conway, Peter Fenwick, Brian Goodwin, Susan Greenfield, Valerie Hall, Jane Henry, Tony Marcel, Richard Morris, Brian Nicholson, Dick Passingham, Mark Price, Robin Skynner, John Teasdale, Francisco Varela, Max Velmans and Larry Weiskrantz. Special thanks to Margaret Carr for wonderful conversations, and to her and her husband Malcolm for their friendship, and the use, yet again, of their muse-filled beach-house in Raglan, New Zealand. Michelle Macdonald, Steven Smith and Christopher Titmuss helped me to practise the art of thinking slowly. And Christopher Potter and Emma Rhind-Tutt of Fourth Estate believed in the book enough to chivvy me into improving it. Emma's love of language and her persistent refusal to let me get away with sloppiness of mind or prose helped enormously to shape the book for the better. Such inaccuracies and infelicities as remain are, of course, down to me.

I am grateful to the following authors and publishers for permission to reproduce quotations and illustrations.

Academic Press Inc., and Professor Patricia Bowers (executor of the estate of the late Professor Kenneth S. Bowers), for two panels from figure 2, p83 in 'Intuition in the context of discovery', by K.S. Bowers, G. Regehr, C. Balthazard and K. Parker, reprinted with kind permission from *Cognitive Psychology*, vol 22, pp72–110, © 1990 Academic Press.

American Psychological Association and Professor J. Schooler for illustrations from the appendix (p182) to J.W. Schooler, S. Ohlsson and K. Brooks (1993) 'Thoughts beyond words: when language overshadows insight', *Journal of Experimental Psychology: General*, vol 122, pp166–183; © 1993 by the American Psychological Association. Reprinted with permission.

The British Psychological Society for extracts from the report 'Fostering Innovation: A Psychological Perspective' by M.A. West, C. Fletcher and J. Toplis, March 1994.

Cambridge University Press for extracts from A.E. Housman, *The Name and Nature of Poetry.*

Elsevier Science Ltd for the extract from 'How does cognitive therapy prevent depressive relapse and why should attentional control (mindfulness) training help?' by John Teasdale, Zindel Segal and Mark Williams, reprinted from *Behavior Research and Therapy*, 1995, vol 33, pp25–39. © 1995, with kind permission from Elsevier Science Ltd, The Boulevard, Langford Lane, Kidlington, OX5 1GB, UK.

Faber and Faber Ltd. for the extract from Ted Hughes, *Poetry in the Making*, published by Faber and Faber Ltd., London 1967. © Ted Hughes 1967, reprinted with kind permission.

HarperCollins Publishers for extracts from *Discourse on Thinking*, translated by John M. Anderson and E. Hans Freund by Martin Heidegger. Copyright © 1959 by Verlag Gunther Neske. Copyright © in the English Translation by Harper & Row, Publishers Inc. Reprinted by permission of HarperCollins Publishers, Inc. Excerpts from *Women's Ways of Knowing* by Mary Field Belenky et al. Copyright © 1986 by Basic Books, Inc. Reprinted by permission of BasicBooks, a division of HarperCollins Publishers, Inc.

The MIT Press for the extract from P.S. Churchland. *Neurophilosophy*, MIT Press, Cambridge, MA. © 1986 Patricia S. Churchland.

Oxford University Press for the extract from Arthur Reber, *Implicit Learning and Tacit Knowledge: An Essay on the Cognitive Unconscious*, OUP, Oxford, 1993.

Princeton University Press for the extract from D.T. Suzuki, *Zen and Japanese Culture*, Princeton University Press, 1959; and for the extract from J. Hademard, *The Psychology of Invention in the Mathematical Field*, Princeton University Press, 1945.

Random House Inc. for the excerpt from F. Scott Fitzgerald, *Tender is the Night*, published by The Bodley Head; and for the excerpts from Tom Peters, *The Pursuit of Wow!*, published by Vintage Books.

Scientific American Inc., New York, for permission to reproduce the schematic representation of a neuron first published in 'The chemistry of the brain' by Leslie L. Iversen, in *Scientific American*, September, 1979, and later in *The Brain: A Scientific American Book*.

Simon and Schuster Inc. for the extract from Richard Selzer, *Mortal Lessons*, © 1974, 1975, 1976.

John Wiley and Sons Ltd. for the extract from Nicholas Humphrey, in discussion of Chapter by John Kihlstrom, in CIBA Symposium 174, *Experimental and Theoretical Studies of Consciousness*, Wiley, Chichester, 1993.

Figures

CHAPTER 1

The Speed of Thought

Turtle buries its thoughts, like its eggs, in the sand, and allows the sun to hatch the little ones. Look at the old fable of the tortoise and the hare, and decide for yourself whether or not you would like to align with Turtle.

Native American Medicine Cards

There is an old Polish saying, 'Sleep faster; we need the pillows', which reminds us that there are some activities which just will not be rushed. They take the time that they take. If you are late for a meeting, you can hurry. If the roast potatoes are slow to brown, you can turn up the oven. But if you try to speed up the baking of meringues, they burn. If you are impatient with the mayonnaise and add the oil too quickly, it curdles. If you start tugging with frustration on a tangled fishing line, the knot just becomes tighter.

The mind, too, works at different speeds. Some of its functions are performed at lightning speed; others take seconds, minutes, hours, days or even years to complete their course. Some can be speeded up – we can become quicker at solving crossword puzzles or doing mental arithmetic. But others cannot be rushed, and if they are, then they will break down, like the mayonnaise, or get tangled up, like the fishing line. 'Think fast; we need the results' may sometimes be as absurd a notion, or at least as counterproductive, as the attempt to cram a night's rest into half the time. We learn, think and know in a variety of different ways, and these modes of the mind operate at different speeds, and are good for different mental jobs. 'He who hesitates is lost', says one proverb. 'Look before you leap', says another. And both are true.

Roughly speaking, the mind possesses three different processing speeds. The first is faster than thought. Some situations demand an unselfconscious, instantaneous reaction. When my motor-bike skidded on a wet manhole cover in London some years ago, my brain and my body immediately choreographed for me an intricate and effective set of movements that enabled me to keep my seat – and it was only after the action was all over that my conscious mind and my emotions started to catch up. Neither a concert pianist nor an Olympic fencer has time to figure out what to do next. There is a kind of 'intelligence' that works more rapidly than thinking. This mode of fast, physical intelligence could be called our 'wits'. (The five senses were originally known as 'the five wits'.)

Then there is thought itself: the sort of intelligence which does involve figuring matters out, weighing up the pros and cons, constructing arguments and solving problems. A mechanic working out why an engine will not fire, a family arguing over the brochures about where to go for next summer's holiday, a scientist trying to interpret an intriguing experimental result, a student wrestling with an examination question: all are employing a way of knowing that relies on reason and logic, on deliberate conscious thinking. We often call this kind of intelligence 'intellect' – though to make the idea more precise, I shall call it *d-mode*, where the 'd' stands for 'deliberation'. Someone who is good at solving these sorts of problems we call 'bright' or 'clever'.

But below this, there is another mental register that proceeds more slowly still. It is often less purposeful and clear-cut, more playful, leisurely or dreamy. In this mode we are ruminating or mulling things over; being contemplative or meditative. We may be pondering a problem, rather than earnestly trying to solve it, or just idly watching the world go by. What is going on in the mind may be quite fragmentary. What we are thinking may not make sense. We may even not be aware of much at all. As the English yokel is reported to have said: 'sometimes I sits and thinks, but mostly I just sits'. Perched on a seaside rock, lost in the sound and the motion of the surf, or hovering just on the brink of sleep or waking, we are in a different mental mode from the one we find ourselves in as we plan a meal or dictate a letter. These leisurely, apparently aimless, ways of knowing and experiencing are just as 'intelligent' as the other, faster ones. Allowing the mind time to meander is not a luxury that can safely be cut back as life or work gets more demanding. On the contrary, thinking slowly is a vital part of the cognitive armamentarium. We need the tortoise mind just as much as we need the hare brain.

Some kinds of everyday predicament are better, more effectively approached with a slow mind. Some mysteries can *only* be penetrated with a relaxed, unquesting mental attitude. Some kinds of understanding simply refuse to come when they are called. As the *Tao Te Ching* puts it:

> Truth waits for eyes unclouded by longing.
> Those who are bound by desire see only the outward container.

Recent scientific evidence shows convincingly that the more patient, less deliberate modes of mind are particularly suited to making sense of situations that are intricate, shadowy or ill defined. Deliberate thinking, d-mode, works well when the problem it is facing is easily conceptualised. When we are trying to decide where to spend our holidays, it may well be perfectly obvious what the parameters are: how much we can afford, when we can get away, what kinds of things we enjoy doing, and so on. But when we are not sure what needs to be taken into account, or even which questions to pose – or when the issue is too subtle to be captured by the familiar categories of conscious thought – we need recourse to the tortoise mind. If the problem is not whether to go to Turkey or Greece, but how best to manage a difficult group of people at work, or whether to give up being a manager completely and retrain as a teacher, we may be better advised to sit quietly and ponder than to search frantically for explanations and solutions. This third type of intelligence is associated with what we call creativity, or even 'wisdom'.

Poets have always known the limitations of conscious, deliberate thinking, and have sought to cultivate these slower, mistier ways of knowing. Philosophers from Spinoza and Leibniz to Martin Heidegger and Suzanne Langer have written about the realms of mind that lie beyond and beneath the conscious intellect. Psychotherapists know that 'the unconscious' is not just a source of personal difficulties; a revised *relationship* with one's unconscious is also part of the 'cure'. And the sages and mystics of all religious traditions attest to the spontaneous transformation of experience that occurs when one embraces the 'impersonal mystery' at the core of mental life – whether this mystery be the 'godhead' of Meister Eckhart or the 'Unborn' of Zen master Bankei. Even scientists themselves, or at least the most creative of them, admit that their genius comes to them from layers of mind over which they have little or no control (and they may even feel somehow fraudulent for taking personal credit for insights that simply 'occurred to them').[1]

It is only recently, however, that scientists have started to explore the slower, less deliberate ways of knowing directly. The newly formed hybrid discipline of 'cognitive science', an alliance of neuroscience, philosophy, artificial intelligence and experimental psychology, is revealing that the unconscious realms of the human mind will successfully accomplish a number of unusual, interesting and important tasks *if they are given the time.* They will learn patterns of a degree of subtlety which normal consciousness cannot even see; make sense out of situations that are too complex to analyse; and get to the bottom of certain difficult issues much more successfully than the questing intellect. They will detect and respond to meanings, in poetry and art, as well as in relationships, that cannot be clearly articulated.

One of my main aims in writing this book is to bring this fascinating research to a wider audience, for it offers a profound and salutary challenge to our everyday view of our own minds and how they work. These empirical demonstrations are more than interesting: they are important. For my argument is not just that the slow ways of knowing exist, and are useful. It is that our culture has come to ignore and undervalue them, to treat them as marginal or merely recreational, and in so doing has foreclosed on areas of our psychological resources that we need. Just like the computer, the Western mind has come to adopt as its 'default mode' just one of its possible modes of knowing: d-mode. (The 'd' can stand for 'default' as well as 'deliberation'.)

The individuals and societies of the West have rather lost touch with the value of contemplation. Only active thinking is regarded as productive. Sitting gazing absently at your office wall or out of the classroom window is not of value. Yet many of those whom our society admires as icons of creativity and wisdom have spent much of their time doing nothing. Einstein, it is said, would frequently be found in his office at Princeton staring into space. The Dalai Lama spends hours each day in meditation. Even that paragon of penetrating insight, Sherlock Holmes, is described by his creator as entering a meditative state 'with a dreamy vacant expression in his eyes'.

There are a number of reasons why slow knowing has fallen into disuse. Partly it is due to our changing conception of, and attitude towards, time. In pre-seventeenth-century Europe a leisurely approach to thinking was much more common, and in other cultures it still is. A tribal meeting at a Maori *marae* can last for days, until everyone has had time to assimilate the issues, to have their say, and to form a consensus. However, the idea that time is plentiful is in many parts of the world now seen as laughably old-fashioned and self-indulgent.

Swedish anthropologist Helena Norberg-Hodge has documented

the way in which the introduction of Western culture has radically altered the pace of life in the traditional society of Ladakh, for example.[2] Until ten years ago, a Ladakhi wedding lasted a fortnight. But their lifestyle rapidly altered following the introduction of some simple 'labour-saving' changes: tools, such as the Rotovator, to make ploughing quicker and easier; and some new crops and livestock, such as dairy cows. Compared to the traditional yak, cows yield more milk than a family needs, creating a surplus which can be turned into cheese and sold to bring in some extra cash. While there is no harm in making life a little easier, in encouraging families to accumulate a little 'wealth', unfortunately this apparently benign 'aid package' also gave the Ladakhis a new view of time – as something in short supply. Instead of the Rotovators and the cows generating *more* leisure, they have in fact reduced it. People are now busier than they were: busy creating wealth – and 'saving time'. Today a Ladakhi wedding lasts less than a day, just like an English one. Within the Western mindset, time becomes a commodity, and one inevitable consequence is the urge to 'think faster': to solve problems and make decisions quickly.

Partly the decline of slow thinking is to do with the rise of what the American social critic Neil Postman has called 'technopoly' – the widespread view that every ill is a problem which has a potential solution; solutions are provided by technological advances, which are generated by clear, purposeful, disciplined thinking; and the faster problems are solved, the better. Thus, as the Ladakhis have recently joined us in believing, time is an adversary over which technology can triumph. For Postman, technopoly is based on

> the beliefs that the primary, if not the only goal of human labour and thought is efficiency; that technical calculation is in all respects superior to human judgment; that in fact human judgment cannot be trusted, because it is plagued by laxity, ambiguity, and unnecessary complexity; that subjectivity is an obstacle to clear thinking; that what cannot be measured either does not exist or is of no value; and that the affairs of citizens are best guided and conducted by 'experts.'[3]

In such a culture, time spent exploring the question is only justified to the extent that it clearly leads towards a solution to the problem. To spend time dwelling on the question to see if it may lead to a *deeper* question seems inefficient, self-indulgent or perverse.

In contemporary 'Western' society (which now effectively covers the globe), we seem to have generated an inner, psychological culture of speed, pressure and the need for control – mirroring the outer culture of efficiency and productivity – in which access to the slower modes of mind has been lost. People are in a hurry to know, to have answers, to plan and solve. We urgently want explanations: Theories of Everything, from marital mishaps to the origin of the universe. We want more data, more ideas; we want them faster; and we want them, with just a little thought, to tell us clearly what to do.

We find ourselves in a culture which has lost sight (not least in its education system) of some fundamental distinctions, like those between being wise, being clever, having your 'wits' about you, and being merely well informed. We have been inadvertently trapped in a single mode of mind that is characterised by information-gathering, intellect and impatience, one that requires you to be explicit, articulate, purposeful and to show your reasoning. We are thus committed (and restricted) to those ways of knowing that can function in such a high-speed mental climate: predominantly those that use language (or other symbol systems) as a medium and deliberation as a method. As a culture we are, in consequence, very good at solving analytic and technological problems. The trouble is that we tend, increasingly, to treat all human predicaments as if they were of this type, including those for which such mental tools are inappropriate. We meet with cleverness, focus and deliberation those challenges that can only properly be handled with patience, intuition and relaxation.

To tap into the leisurely ways of knowing, one must dare to wait. Knowing emerges from, and is a response to, not-knowing. Learning – the process of coming to know – emerges from uncertainty. Ambivalently, learning seeks to reduce uncertainty, by transmuting the strange into the familiar, but it also needs to tolerate uncertainty, as the seedbed in which ideas germinate and responses form. If either one of these two aspects of learning predominates, then the balance of the mind is disturbed. If the passive acceptance of not-knowing overwhelms the active search for meaning and control, then one may fall into fatalism and dependency. While if the need for certainty becomes intemperate, undermining the ability to tolerate confusion, then one may develop a vulnerability to demagoguery and dogma, liable to cling to opinions and beliefs that may not fit the bill, but which do assuage the anxiety.

Perhaps the most fundamental cause of the decline of slow knowing, though, is that as a culture we have lost our sense of the *unconscious intelligence* to which these more patient modes of mind give access, a loss for which René Descartes conventionally takes the blame. If the busy conscious mind is to allow itself to wait, mute, for something to come, presumably from a source beyond its ken and its control, it has, minimally, to acknowledge the existence of such a source. Modern Western culture has so neglected the intelligent unconscious – the *undermind*, I shall sometimes call it – that we no longer know that we have it, do not remember what it is for, and so cannot find it when we need it.[4] We do not think of the unconscious as a valuable resource, but (if we think of it at all) as a wild and unruly 'thing' that threatens our reason and control, and lives in the dangerous Freudian dungeon of the mind.[5] Instead, we give exclusive credence to conscious, deliberate, purposeful thinking – d-mode. Broader than strict logic or scientific reasoning, though it includes these, d-mode has a number of different facets.

D-mode is much more interested in finding answers and solutions than in examining the questions. Being the primary instrument of technopoly, and as such centrally concerned with problem-solving, d-mode treats any unwanted or inconvenient condition in life as if it were a 'fault' in need of fixing; as if one's loss of libido or turnover were technical malfunctions which one ought – either by oneself, or with the aid of an 'expert', such as a counsellor or a market analyst – to be able to put right.

D-mode treats perception as unproblematic. It assumes that the way it sees the situation is the way it is. The diagnosis is taken for granted. The idea that the fault may be in the way the situation is perceived or 'framed', or that things might look different 'on closer inspection', does not come naturally to d-mode.

D-mode sees conscious, articulate understanding as the essential basis for action, and thought as the essential problem-solving tool. The activity in d-mode is predominantly that of gaining a mental grasp, or figuring out. This may involve the impeccable rationality of the prototypical scientist, with her equations and flow charts and technical terms. Or it may involve the more common-or-garden kinds of thinking: weighing up the pros and cons of a decision; talking things through with a friend; jotting down thoughts or making lists on the back of an envelope; trying out arguments over dinner, discussing family arrangements, making a sales pitch. Though this latter kind of thinking may not match up to the exacting standards of the professional philosopher or mathematician, and is often full of unnoticed holes, nevertheless it is, in its form and intent, 'quasi' or 'proto'-rational.

D-mode values explanation over observation, and is more concerned about 'why' than 'what'. Sometimes figuring out is designed to get directly to the point of action. But commonly, either as a means or an end in itself, what it seeks is understanding or explanation. The need to have a mental grasp, to be able to offer, to oneself if not to others, an acceptable account of things, is an integral part of d-mode. Right from playschool, adults will be asking children: 'What are you trying to do?', or 'That's interesting; why did you do that?' And children quickly get the idea that they ought to know what they are up to, what they are trying to achieve; and to be able to give an account of themselves, their actions and their motives, to other people. They come to assume, with their parents and teachers, that it is normal to be intentional, and proper to have explanations to offer. As ever, there is no problem with this *per se*; it is a very useful ability. But when this purposeful, justificatory, 'always-show-your-reasoning' attitude becomes part of the dominant default mode of the mind, it then tends to suppress other ways of knowing, and makes one sceptical of any activity whose 'point' you cannot immediately, consciously see.

D-mode likes explanations and plans that are 'reasonable' and justifiable, rather than intuitive. The demand that ideas always come with supporting arguments and explanations may lead one to reject out of hand thoughts that are in fact extremely fruitful, but which arrive without any indication of their pedigree or antecedents. The productive intuition can be overlooked in favour of the well-argued case. And if explanation comes to be seen as a necessary intermediary between a problem and a plan of action – if one does not feel qualified to act without a conscious rationale – then again one might miss out on some short cuts and bright ideas. Doubt, in the sense of a lack of conscious comprehension, becomes stultifying rather than facilitating; a trap rather than a springboard.

D-mode seeks and prefers clarity, and neither likes nor values confusion. Because of its concern with justification, d-mode likes to move along a well-lit path from problem to solution, preserving, as it goes, as much of a mental grasp as it can. It prefers learning that hops from stepping-stone to stepping-stone, without getting its feet wet, like a mathematical proof, or a well-argued report that progresses smoothly from a problem, to a clear analysis, to a plausible solution, to an action plan. And while some learning may proceed in this point-by-point fashion, much does not. Often learning emerges in a more gradual, holistic way, only after a period of casting around for a vague sense of direction, like a pack of hounds that has lost the scent. An artist composing a still life, a client in psychotherapy, even a scientist on the verge of a breakthrough: none of these (as we shall see) would be functioning optimally in d-mode. To undertake this kind of slow learning, one needs to be able to feel comfortable being 'at sea' for a while.

D-mode operates with a sense of urgency and impatience. It is accompanied by a subtle – or sometimes gross – sense of not having enough time; of wanting things to be sorted out soon; of getting irritable when the fix is not quick enough. Fuelled by this sense of urgency, we find ourselves living, increasingly, in the fast lane. And the technology – be it planes or Powerbooks, microwaves or modems – tracks this need, but also channels and exacerbates it. If you have to wait for the TV news, or tomorrow's newspapers, to hear about the rumours on Wall Street, or a small earthquake in Peru, you're not a serious player. Our intolerance of dissatisfaction, or even of a delay in information, comes to dictate the kind of mind-mode with which we meet *any* kind of adversity.

D-mode is purposeful and effortful rather than playful. With problem-solving and impatience comes a feeling of mental strain, of pushing for answers that would not arrive by themselves, or certainly not quickly enough. In d-mode there is always this sense, vague or acute, of being under time pressure, and of being intentional, purposeful, questing: of needing to have an answer to a pre-existing question, whether it concerns a fault in the production line or the meaning of life. Once this busy activity becomes all we know how to do, the default mode, then we are going to miss any fruits of *relaxed cognition.*

D-mode is precise; it tends to work with propositions made up of clearly defined symbols, preferably the hyper-precise languages of mathematics and science, where every term seems to be transparent and complete. A model of the national economy which can be represented as a sophisticated computer program, in which everything that counts can be given a measure – and in which therefore everything which has no measure has no place and no value – is taken more seriously than one which may subsume a richer view of human nature, but which is less explicit and precise. The history of scientific psychology – a d-mode enterprise if ever there was one – is full of precise theories about how memory works, for example, which make quantitative predictions about arcane laboratory tasks, but which simply ignore almost everything that people find interesting about their own powers of retention. When I was working on memory for my doctorate, I stopped telling people at parties because they would inevitably start to ask me all kinds of fascinating questions to which my detailed knowledge was embarrassingly irrelevant. (Happily things in memory research have improved somewhat in the last twenty-five years.)

D-mode relies on language that appears to be literal and explicit, and tends to be suspicious of what it sees as the slippery, evocative world of metaphor and imagery. If something can be understood, it can be understood clearly and unambiguously, says the intellect. An intimation of understanding that does not quite reveal itself, that remains shrouded or indistinct, is, to d-mode, only an impoverished kind of understanding; one that should either be forced to explain itself more fully, or treated with disdain. Poetry does not capture anything that cannot ultimately be better, more clearly rendered in prose, and rhetoric is a poor cousin of reasoned explanation.

D-mode works with concepts and generalizations, and likes to apply 'rules' and 'principles' where possible. D-mode favours abstraction over particularity. It works with what is generic or prototypical. It talks about 'the workforce', 'the rational consumer', 'the typical teacher', 'the environment', 'holidays', 'feelings'. Even individuals are treated as generalizations, collections of traits and dispositions. 'John Major' and 'Cher' are as much abstractions as 'the national debt' or 'the state of Welsh rugby'. The idea that a kind of truth could be derived from a close, sustained but unthinking attention to a single object is foreign to d-mode.

Language necessarily imposes a certain speed, a particular time frame, on cognition, so *d-mode must operate at the rates at which language can be received, produced and processed.* If you speed speech up it soon becomes unintelligible. If you slow it down beyond a certain point it loses its meaning. (Old-fashioned vinyl '45s', when played at either 33 or 78 revolutions per minute, demonstrate this phenomenon nicely.) Those modes of mind that work very slowly (or, for that matter, very fast) cannot, therefore, operate with the familiar tools of words and sentences. They need different contents, different elements – or perhaps no conscious elements at all. And without the familiar ticker tape of words rolling across the screen of consciousness, there may come a disconcerting feeling of having lost predictability and control. Thus *d-mode maintains a sense of thinking as being controlled and deliberate*, rather than spontaneous or wilful.

D-mode works well when tackling problems which can be treated as an assemblage of nameable parts. It is in the nature of language to segment and analyse. The world seen through language is one that is perforated, capable of being gently pulled apart into concepts that seem, for the most part, self-evidently 'real' or 'natural', and which can be analysed in terms of the relationships between these concepts. Much of traditional science works so well precisely because the world of which it treats is this kind of world. But when the mind turns its attention to situations that are ecological or 'systemic', too intricate to be decomposed in this way without serious misrepresentation, the limitations of d-mode's linguistic, analytical approach are quickly reached. Any situation that is organic rather than mechanical is likely to be of this kind. The new 'sciences' of chaos and complexity are in part a response to the realisation that d-mode is *in principle* unequal to the task of explaining systems as complicated as the weather, or the behaviour of animals in the natural world. Along with the rise of these new sciences must come a re-evaluation of the slower ways of knowing; of intuition as an essential complement to reason.

The fact that language can handle only so much complexity is easy to demonstrate. Take the sentence

> The ecologist hated the accountant.

This is trivially easy to understand. Now take

> The accountant the ecologist hated abused the waiter.

This is still perfectly manageable. Add another (quite grammatical) embedded clause

> The waiter the accountant the ecologist hated abused loved the archbishop.

Understanding begins to get slightly tenuous. And when we get to

> The archbishop the waiter the accountant the ecologist hated abused loved joined the conspiracy

you have to work quite hard. You begin to need some kind of cognitive prosthesis, like a diagram, if you are to overcome the limitations of memory and understanding that are being revealed. Without the build-up, it would take some very deliberate unpacking to figure out who it was who abused whom. D-mode stretched to its limit becomes cumbersome and inept.

Here are two other examples of perfectly grammatical language that are, in practice, virtually incomprehensible.

> We cannot prove the statement which is arrived at by substituting for the variable in the statement form 'We cannot prove the statement which is arrived at by substituting for the variable

> in the statement form Y the name of the statement form in question', the name of the statement form in question.

And:

> Both is preferable to neither; but naturally both both and neither is preferable to neither both nor neither; but naturally both both both and neither and neither both nor neither is preferable to neither both both and neither nor neither both nor neither; but – naturally – both both both both and neither and neither both nor neither and neither both both and neither nor neither both nor neither is preferable to neither both both both and neither and neither both nor neither nor neither both both and neither nor neither both nor neither.

Unless we have spent years getting used to statements like this, d-mode simply has to give up. A professor of logic might be able to make her way through these abstract jungles, but the fact that d-mode admits of levels of expertise should not blind us to its inherent limitations. Even language and logic can rapidly get out of control if we let them. And it is therefore an open question whether there are kinds and degrees of complexity which might be handled better in a different way.

If we see d-mode as the only form of intelligence, we must suppose, when it fails, that we are not 'bright' enough, or did not think 'hard' enough, or have not got enough 'data'. The lesson we learn from such failures is that we must develop better models, collect more data, and ponder more carefully. What we do not learn is that we may have been thinking *in the wrong way*. While this epistemological stance remains invisible and unchallenged, therefore, the search for better answers to personal, social, political and environmental predicaments has to be conducted by the light of conscious thought. Our efforts are like those of the man who was searching for his car keys under the streetlight – though he has lost them elsewhere – because that was the only place he could *see*. Thus scientists, researchers, intellectuals and those who program computers with complicated formulae in order to try to predict economic trends remain the people on whom we tend to pin our hopes in the face of difficulties and uncertainties. They are the ones who, by general acclaim, have the best, most explicit models; who have the most information; and who are the most skilled thinkers. We trust them. Where else could we look for guidance?

The 'slow ways of knowing' are, in general, those that lack any or all of the characteristics of d-mode. They spend time on uncovering what may lie behind a particular question. They do not rush into conceptualisation, but are content to explore more fully the situation itself before deciding what to make of it. They like to stay close to the particular. They are tolerant of information that is faint, fleeting, ephemeral, marginal or ambiguous; they like to dwell on details which do not 'fit' or immediately make sense. They are relaxed, leisurely and playful; willing to explore without knowing what they are looking for. They see ignorance and confusion as the ground from which understanding may spring. They use the rich, allusive media of imagination, myth and dream. They are receptive rather than proactive. They are happy to relinquish the sense of control over the directions that the mind spontaneously takes. And they are prepared to take seriously ideas that come 'out of the blue', without any ready-made train of rational thought to justify them. These are the modes of mind that the following chapters will explore, in order both to reveal their nature and their value and also to uncover ways in which they might be rehabilitated.

In order to rehabilitate the slow ways of knowing, we need to adopt a different view of the mind as a whole: one which embraces sources of knowledge that are less articulate, less conscious and less predictable. The undermind is the key resource on which slow knowing draws, so we need new metaphors and images for the relationship between conscious and unconscious which escape from the polarisation to which both Descartes and Freud, from their different sides, subscribed. Only in the light of new models of the mind will we see the possibility and the point of more patient, receptive ways of knowing, and be able to cultivate – and tolerate – the conditions which they require.

The crucial step in this recovery is not the acquisition of a new psychological technology (brainstorming, visualisation, mnemonics and so on), but a revised understanding of the human mind, and a willingness to move into, and to enjoy, the life of the mind as it is lived in the shadowlands rather than under the bright lights of consciousness. Clever mental techniques – devices that 'tap' the resources of the 'right hemisphere', as if it were a barrel of beer – miss the point if they leave in place the same questing, restless attitude of mind. In many courses on 'creative management' or 'experiential learning', it is a case of *plus ça change, plus c'est la même chose.* Instead of calling a meeting to 'discuss' the problem, you call one to 'brainstorm' it, or to get people to draw it with crayons. But the pressure for results, the underlying impatience, is still there. The key to the undermind is not an overlay of technique but radical reconceptualisation. When the mind slows and relaxes, other ways of knowing automatically reappear. If and when this shift of mental mood takes place, *then* some different strategies of thought may indeed be helpful, but, without it, they are useless. (This, incidentally, explains why the enthusiasm for each new, much-hyped mental technology has such a disappointingly short half-life.)[6]

Another step in the recovery of the slower ways of knowing is to recognise that these forms of cognition are not the exclusive province of special groups of people – poets, mystics or sages – nor do they appear only on special occasions. They have sometimes been talked about in rather mystifying ways, as the work of 'the muse', or as signifying great gifts, or special states of grace. Such talk makes slow knowing look rather awesome and arcane. One feels intimidated, as if such mental modes were beyond the reach of ordinary mortals, or had little to do with the mundane realities of modern life. This is a false and unhelpful impression. A 'poetic way of knowing' is not the special prerogative of those who string words together in certain ways. It is accessible, and of value, to anyone. And though it cannot be trained, taught or engineered, it can be cultivated by anyone.

So *Hare Brain, Tortoise Mind* is about why it is a good idea to pull off the Information Super-Highway into the Information Super Lay-By; to stop chasing after more data and better solutions and to rest for a while. It is about why it is sometimes more intelligent to be less busy; why there are ideas one can gain access to by loafing which are inaccessible to earnest, purposeful cognition. And it is about the reasons why these natural endowments of the human mind have become neglected in twentieth-century Euro-American culture, and why, in this culture, they are sorely missed.

CHAPTER 2

Basic Intelligence: Learning by Osmosis

It is a profoundly erroneous truism, repeated by all copybooks and by eminent people when they are making speeches, that we should cultivate the habit of thinking of what we are doing. The precise opposite is the case. Civilization advances by extending the number of important operations which we can perform *without* thinking about them. Operations of thought are like cavalry charges in battle – they are strictly limited in number, they require fresh horses, and must only be made at decisive moments.

A. N. Whitehead

It is February: summer in New Zealand. I am closeted with my laptop in a beach-house on the west coast of the North Island (overlooking what surfers tell me is the best left-hand break in the Southern Hemisphere), and there are a lot of flies. I find them, especially the big brown ones, very distracting, so, despite my Buddhist leanings, I swat them. There are also a number of spiders, long-legged and small-bodied, which I rather like. This morning I dropped a freshly swatted fly into the web of one of the spiders. I then proceeded to watch, rapt, for twenty minutes as the spider manoeuvred the fly from where I had placed it to its own dining area, a distance of some twelve centimetres. First it spun a coat round the fly to hold it secure. Then it delicately cut away the strands of web that were supporting it, until it was left dangling only by a few threads. Holding on to the web above with two legs, and clasping the fly with the others, it pulled it towards its destination by about half a centimetre, and secured it with another thread. It released the strands that were now holding the fly back, allowing it to swing a little towards the goal, and spun some more ties to hold it in its new position. Taking up the diagonal position, it hauled its prize sideways again, secured it, and then cast off the ropes that were restraining it. And so it went, until lunch was finally in the right place.

The equivalent task for me, I computed, would have been something like single-handedly transporting a blue whale a distance of 120 metres across a bottomless abyss, equipped only with some very strong elastic, grappling hooks, and a sharp knife. This perilous feat of engineering would have taken a great deal of thought and calculation to counter the constant risk that one false move, such as cutting the wrong string at the wrong moment, would send the whale, and very possibly me as well, plummeting into the void. The spider, whose whole body was two millimetres long, with a minute brain, didn't make a mistake. I was impressed. I did not feel obliged to credit the spider with consciousness; but I had to marvel at its intelligence.

There is a resurgence of interest in the concept of 'intelligence' these days, prompted, in large measure, by a growing dissatisfaction with the assumption that d-mode is the be-all and end-all of human cognition. Harvard psychologist Howard Gardner has suggested that there are 'multiple intelligences', of which he claims to have identified eight and a half, and which resemble quite closely the subjects of the traditional school curriculum.[1] Daniel Goleman argues for a rapprochement between reason and feeling with his notion of 'emotional intelligence'.[2] But to understand more broadly how the different facets of intelligence fit together, we have to find an approach that does not presuppose the primacy of the intellect.

At its most basic, intelligence is what enables an organism to pursue its goals and interests as successfully as possible in the whole intricate predicament in which it finds itself. My spider had been designed by evolution to perform, within its own world, the most challenging of tasks in an efficient and sophisticated way. These miracles of intelligent adaptation are commonplace in the animal kingdom, and many of them have been documented rather more systematically than my spider. If a rat eats a meal that consists of a mixture of a familiar and an unfamiliar food, and subsequently becomes sick, it will in future avoid the new food but not the familiar one.[3] That, I think, is intelligent.

Much of human intelligence, too, has little to do with d-mode. A baby is being intelligent when it smiles hopefully at its mother, or turns its head away from a looming object. A teenager is being intelligent when learning to get along in school by blending into the background, or deploying a disarming humour. A poet is being intelligent when considering a variety of candidates for the *mot juste.* And though a mathematician is also being intelligent as she tries to work out the solution to a complex problem, her finely honed intellectual ability is just one variety of intelligence, and a rather peculiar and arcane one at that. Intelligence may be associated with words, logical argument, explicit trains of thought or articulate explanation, but it may equally well not be. Fundamentally, intelligence helps animals, including human beings, to survive.

The most basic of these strategies, common to all levels of life from amoebas to archbishops, is the bred-in-the-bone tendency to approach and maintain conditions which favour survival, and to avoid or escape from conditions which are aversive. The former conditions we call 'needs'; the latter 'threats'. Evolution has equipped every animal with a repertoire, large or small, of ways to minimise the risk of damage and enhance its wellbeing. The spider weaves its web, manoeuvres its prey, and freezes when the air moves in a disturbing way. The digger wasp, *sphex ichneumoneus*, which cleverly buries a paralysed cricket alongside her eggs for the newborn grubs to feed on, always leaves the cricket outside her burrow while she goes in to make sure all is well, before dragging it in.[4] Even potential threats are allowed for in such reflex behaviour.

But the genetically given reactions to threats of 'fight, flight, freeze and check', though helpful, are by no means infallible. A spider may go still, despite being dangerously highlighted against the background of a white bathtub. If an interfering ethologist steals the cricket every time *sphex* makes her subterranean safety check, she cannot adapt. She never realises that in this new world it may make more sense to drag the cricket in with her first time. A baby shows distress even though the looming object is actually a rapidly inflated balloon, and not a projectile. Reflexes, though intelligent, may be turned on their heads by unprecedented events: those for which evolution has not had time to prepare you. Such reflexes provide a vital starter kit of survival intelligence, but if the ability to build on these wired-in, entry-level responses is lacking, an animal remains highly vulnerable to change.

So the next stage in the evolution of intelligence is *learning*. Gathering knowledge and developing expertise are survival strategies. In unfamiliar situations, animals are at risk. They are unable to predict and control what is going on. Potential sources of succour may go unrecognised. Actual threats may not be perceived until it is too late. Uncertainty may always conceal danger. The ability to reduce uncertainty, to convert strangeness into known-ness, therefore offers a powerful evolutionary advantage. All the different ways of learning and knowing which human beings possess, however sophisticated, spring ultimately from this biological imperative. Crudely, we might say that 'knowing' is a state in which useful patterns in the world have been registered, and can be used to guide future action. 'Learning' is the activity whereby these patterns are detected. And, at this level, 'intelligence' refers to the resources that make learning, and therefore knowing, possible.

This ability to detect, register and make use of the patterns of relationships that happen to characterise your particular environment is widespread in the animal world. Take the gobiid fish, for example. It has been shown that certain of these fish can find their way from one rock-pool to another, at low tide, by jumping accurately across the exposed rocks. Jumping in this way is a high-risk exercise; if they get it wrong the fish are stranded or injured. In fact, they do it without error. Studies of these fish have ruled out the possibility that they are using some sensory clues such as reflections or smells. If they are placed in an unfamiliar pool, they will not jump. The only possible explanation for this remarkable ability is that, during high tide, the fish swim over and around the crevices and hollows in the sea floor, forming a detailed map of the area which is stored in their memories and used as the basis on which to compute their jumps when they are 'imprisoned' in the low-tide pools.[5]

In the same way, the baby soon comes to know not just the difference between a ball, a balloon and a face, but between her mother's and her father's faces, and tunes her responses accordingly. Her brain is malleable: it is formed not just by the experience of her ancestors, but is also moulded – like that of the gobiid fish, but enormously more so – by the idiosyncrasies of her own experience. A brain is plastic: it transmutes ignorance into competence, and is extraordinarily adept at doing so. Categories and concepts are distilled from particular encounters so that, by a process of spontaneous analogy, 'what I do next' can be informed by records of 'what happened before'. Past mistakes can be avoided and new mistakes made, until, with luck, an effective way of dealing with 'this type of thing' – a big dog, a puncture, an angry face, a new teacher – emerges and confidence is restored. Coming to know the world in this way, to register its patterns and to develop and coordinate skilful responses, is what a sophisticated nervous system – what I shall call a 'brain-mind' – does. It is built to tune itself to certain wavebands of information, and to coordinate these with its own expanding range of capabilities.

After plasticity, the next great development in the evolution of knowing is *curiosity*. Instead of learning simply by reacting to uncertainty, animals became proactive – inquisitive, adventurous, playful. When no more urgent need is occupying your attention, it pays to extend your knowledge, and hence your competence and your security, by going out and actively exploring. So useful is this that evolution has installed it in many species as one of their basic drives. Rats who are allowed to become thoroughly familiar with a maze will quickly explore a new section that is added to it, even though they are being consistently and adequately fed elsewhere. Monkeys kept in a box will repeatedly push open a heavy door to see what is going on outside, and will spend hours fiddling with mechanical puzzles even though they receive no reward for doing so. Human beings who have volunteered to take part in a 'sensory deprivation' experiment, in which all they have to do, to earn forty dollars a day, is to stay in a room with no stimulation, rapidly become desperate for something – anything – to feed their minds, and will repeatedly press a button to hear a voice reading out-of-date stock market quotations.[6]

Being receptive, attentive and experimental, seeking to expand competence and reduce uncertainty, are the design functions of a plastic and enquiring brain-mind. No added encouragement or discipline – no conscious intention, no effort, no deliberation, no articulation – is needed to fulfil this brilliant function. The original design specification of learning does not include the production of conscious rationales. Knowing, at root, is implicit, practical, intuitive. The brain discovers patterns and tunes responses, it is programmed by experience, but this programming is recorded in millions of minute functional changes in the neurons, and manifested in the way the whole organism behaves.

Given the evolutionary priority of this unconscious intelligence – the primacy of *know-how* over *knowledge* – what would we expect the main differences between unconscious and conscious ways of knowing to be?[7] First, we might expect the unconscious to be more robust and resilient, more resistant to disruption, than our conscious abilities. This is exactly what neurological studies of brain damage reveal. When memory, perception or the control of action are degraded, it is the conscious aspects that tend to be lost first, while abilities that are managed unconsciously are spared.[8]

If unconscious abilities are more primitive, more a function of evolution than of culture, we might also expect them to vary less from person to person than do their conscious deliberations. In particular, we should not expect intuitive know-how to show much correlation with measures of 'conscious intelligence' such as IQ – and it doesn't. People's ability to pick up the skills that their everyday lives require – their 'practical intelligence', as Harvard psychologist Robert Sternberg calls it – is independent of their intellectual or linguistic facility. Brazilian street children are able to do the mental arithmetic that their businesses require – quite complex sums, by school standards – without error, despite having very low mathematical ability as measured by tests. People who work as handicappers at American racecourses are able to make calculations, based on a highly intricate model involving as many as seven different variables, yet their ability to do so is completely unrelated to their IQ score.[9]

The minds of children, being immature, must rely more on unconscious than conscious operations. Babies learn to recognise the important people in their lives, and to take an increasingly sophisticated part in the rituals of family life – bathtimes, mealtimes, bedtimes and so on – long before they are able to comment or reflect on what they are doing. They learn to walk by a vast amount of trial and error, out of which is gradually distilled thousands of inarticulate correspondences between the muscles of shoulders, torso and legs, and the sensations of vision, touch and balance. They learn to speak by picking up the language of their culture without any explicit knowledge of grammar. They develop styles of relating without recourse to any instruction book. And as they get older they will learn to ride bicycles, play violins, kick balls, take part in meetings, prepare meals, shop, catch planes and make love, for the most part without being able to explain how it is that they do what they do, or how they learnt it.

The greater part of the useful understanding we acquire throughout life is not explicit knowledge, but implicit know-how. Our fundamental priority is not to be able to talk about what we are doing, but to do it – competently, effortlessly, and largely unconsciously and unreflectingly. And the corresponding need for the kind of learning that delivers know-how – which I shall call *learning by osmosis* – is not one that we outgrow. The brain-mind's ability to detect subtle regularities in experience, and to use them as a guide to the development and deployment of effective action, is our biological birthright. The evolution of more sophisticated strategies complements this basic capability; it does not supersede it. Although the presence of unconscious intelligence is much more obvious in animals and small children, not being overlaid by their conscious, articulate intellect, it is a mistake to suppose that we grow out of it as we get older.

Yet this mistake is made, and it is partly the fault of the renowned Swiss developmental psychologist (or, as he preferred, 'genetic epistemologist') Jean Piaget. Piaget called this ability to master the intuitive craft of living 'sensorimotor intelligence', and claimed that it was of pre-eminent importance during the first two years of life, but was subsequently overtaken and transformed by other more powerful, abstract and increasingly intellectual ways of knowing. In his tremendously influential 'stage theory' of development, Piaget implicitly accepted the cultural assumption that d-mode was the highest form of intelligence, and through the impact that his approach has had on several generations of educators, he inadvertently made sure that schools, even primary schools and kindergartens, saw their job as weaning children off their reliance on their senses and their intuition, and encouraging them to become deliberators and explainers as fast as possible.

The ability to distil out of our everyday experience useful maps and models of the world around us is very down-to-earth; so mundane that it is, in many ways, the unsung hero of the cognitive repertoire. We do this so continuously, so automatically and so unconsciously that it is very easy to overlook just how valuable, and how 'intelligent,' this ability is. It represents the 'poor bloody infantry' of the mind: much less glamorous than the flamboyant cavalry charges of conscious thought. Yet we ignore or disparage this constant honing or sharpening of our 'wits' (in the practical sensory sense of 'wits' that I used in the previous chapter) at our peril, for it turns out that there are things we can learn through this gradual, tacit process which d-mode cannot master; and also that d-mode, if used over-enthusiastically, can actively interfere with this way of knowing. The conscious human intellect stands on the shoulders of learning by osmosis. D-mode is an evolutionary and cultural parvenu, and we cannot properly reassess either its nature or its limitations without looking at its evolutionary underpinnings.

We continue throughout our lives to make use of this unsung ability to pick up patterns and tune our actions accordingly, without being able to say what we have learnt, or even, very often, that we have learnt anything at all. When you start to listen to the works of a particular composer, your mind begins to detect all kinds of characteristics of instrumentation, harmony, rhythm and so on, which enable you to say, on hearing a new piece, 'Isn't that

Bruckner?' Yet how you can tell, unless you are a music scholar, you may be quite unable to say. People who read a lot of whodunnits become, often unconsciously, so familiar with the genre that they know, without thinking, that the murderer is going to be some incidental character in chapter two. When we take a new job, we may consciously collect as much low-down on colleagues-to-be, and the ethos of the workplace, as we can; yet during the first few days and weeks we are also learning a tremendous amount quite automatically: how people greet each other in the morning; how to look busy when we aren't; what kinds of jokes are 'funny' and what are 'crude' or 'sexist'; and so on. As people gain promotions, form stable relationships, have children and are faced with bereavement, so the usefulness of this ability to soak up know-how through their pores does not diminish.

Recent research by psychologists in both Britain and America has reaffirmed the importance of this implicit learning, and shown how it gradually develops over time. Take a professional problem such as learning to regulate the flow of traffic in a city by adjusting the number of buses and the provision and cost of private parking; or to manage a school budget; or to control a complicated industrial process such as the output of a factory or a power station. Situations like this have been studied by Dianne Berry and the late Donald Broadbent at the University of Oxford.[10] Consider the factory production problem. It can be simulated as a 'computer game' in which the levels of various factors, such as the size of the workforce, or of financial incentives, are displayed on the screen, along with the level of the output, and the 'player's' job is to stabilise the output by manipulating the input variables. The effect of each of the variables is actually determined by a reasonably complex equation which the players are not explicitly told.

Players, in their role as 'trainee managers', come, over a period of time, to be able to make adjustments to the input variables that do in fact bring production to the required level – but they are not able to say what they are doing, or explain why it works. When asked to justify a particular 'move', all they may be able to say is that they 'had a hunch', or 'it felt right'. They may even, having made a perfectly good move, say that they thought they were guessing. When the task is quite a difficult one, and people's performance is monitored over several days, their practical know-how and their explicit knowledge – what they can say about their own performance – develop at startlingly different rates. The ability to *do* the job develops relatively quickly, and in some cases quite abruptly; but the ability to articulate that knowledge emerges, if at all, much more slowly.

Broadbent and Berry's laboratory results are by no means unfamiliar in everyday life. Sportspeople and musicians develop high levels of expertise which they are often hard put to analyse or explain. Teachers come to be able to make on-the-spot decisions about how to present a topic or manage a classroom situation, yet may be quite unable to justify their actions to an inquisitive student. In the introduction to a fascinating account of their work on 'principles of problem formation and problem resolution', American psychotherapist Paul Watzlawick and his colleagues describe how the book came about. Working together over several years, they developed some powerful new ways of, as they put it, 'intervening in human problem situations' so as to break through apparent impasses and bring about welcome change. However, as more and more people became interested in their methods through demonstrations and training courses, they became increasingly embarrassed to realise that they had no way of explaining how or why their methods were so successful. 'Only gradually were we ourselves able to conceptualise our approach,' they write – the approach which, at another level, they understood inside out.[11]

Other aspects of this 'implicit learning' have been investigated experimentally by Pawel Lewicki and his colleagues at the University of Tulsa in the United States.[12] Though some of their long series of experiments are rather stylised, they are very illuminating. Like the British researchers, they explored kinds of learning in which people can get better at doing a particular job by picking up subtle patterns embedded in hundreds of examples, but the experimental designs are rather different. In one of these designs, the people taking part in the experiment sit in front of a computer screen which is divided into quarters. Every so often a random-looking array of digits appears on the screen, covering all the four quadrants, and the subjects' job is to detect a particular predetermined number – 6, say – and to push one of four buttons in front of them to indicate which of the four quadrants the 6 is in. The computer automatically records how long it took them to spot the target, and whether their choice of button was right or wrong. There is a brief pause, and then another (different) display appears, and they have to find the 6 again; and so on, for a large number of such 'trials'. The trials are grouped into blocks of seven, with a short break between each block.

As you might expect, the first thing the computer shows is that people get faster at detecting the target as they get more used to,

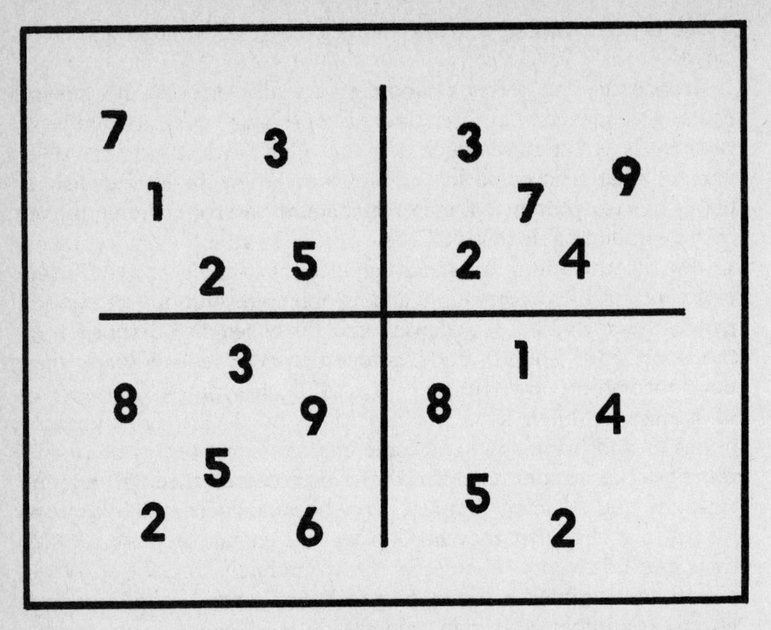

Figure 1. 'In which quadrant is the 6?' – sample grid of numbers used in the Lewicki experiments

and practised at, the task. But now comes the twist. Although it looks to the subjects as if the position of the 6 varies randomly from trial to trial, in fact there is a subtle pattern. Specifically, if you take the positions of the target on the first, third, fourth and sixth trials in a block, you could theoretically predict in which quadrant it is going to appear on the seventh. For example, if the 6 had been in the upper left on trial 1, lower right on trial 3, upper right on trial 4, and lower left on trial 6, then it will appear in the lower right on trial 7. Note that you have to register the positions on *each* of trials 1, 3, 4 and 6, in *each* block of seven. Nothing less than this will give you any useful information at all. Subjects, of course, are not told about this faint pattern. The question is: do they none the less pick it up and make use of it? If they do, this will be shown by the fact that their response times to the seventh target become faster than to the other six. (The general effect of practice and familiarity would obviously not be able to account for this differential effect.)

Sure enough, over a long series of blocks the response to the seventh target becomes progressively quicker than the responses to the other six. Clearly people are picking up the pattern and making use of it. However, when Lewicki showed them the results, all the subjects were surprised at the 'seventh trial' effect – they themselves had not noticed that they were getting selectively faster – and they had no conscious idea what the information might have been that they were, apparently, using. If they were given some more trials, and asked to make a *conscious* prediction for trial 7, they could do no better than chance: 1 in 4.

Lewicki tried very hard to induce in his subjects some conscious awareness of the situation. At the end of several of his studies he told subjects that there *was* a pattern which they had been using, and offered them unlimited time to study all the stimulus arrays, and a sizeable financial reward if they could come up with a suggestion that was close to the actual pattern which he had been using. Nobody was able to say what the pattern was. Next, he ran a group of subjects who were actually his colleagues on the faculty of the University of Tulsa Psychology Department. They should have been able to work out what was going on, if anyone could; they all knew what his research was about. But even they could not consciously detect the pattern. In fact, when they were shown the data proving that they were responding differently to the last trial in each block, some of them confidently accused Lewicki of using subliminal messages to speed them up or slow them down. Now he came to mention it, they said, they had definitely seen something 'fishy' about the displays. Yet there had been nothing fishy at all; merely a pattern that was perfectly visible, if only the conscious mind could have seen it.

The evidence from these studies is clear: we are able unconsciously to detect, learn and use intricate patterns of information which deliberate conscious scrutiny cannot even *see*, under favourable conditions, let alone register and recall. The complexity of Lewicki's patterns (like my impossible-to-understand sentences in the last chapter) was just too great for d-mode to deal with. But when the hare of conscious comprehension ran out of ideas, tortoise mind just kept going. Simply by attending and responding to the situation, without thinking about it, people are able to extract complex patterns of useful information. Of course there are limits to the powers of observation and detection even of the unconscious brain-mind. There must be a great deal of potentially valuable information in the world that is too faint or subtle even for the undermind to detect. But we might, *en passant*, wonder at the wisdom of a society which ignores these unconscious powers, or treats them as ephemera; and of an education system that persists in privileging just one form of conscious, intellectual intelligence over all others.

Mention of education should remind us that even intellectual understanding itself often benefits from this gradual, soaking-it-up-through-the-pores approach. Really 'getting your brain round' a topic seems to depend at least as much on the slower processes of 'mulling over' and 'cogitating' as it does on being mentally busy. Yet many educators seem to be under the impression that people can (and should) master a body of knowledge entirely through d-mode, via intentional study and 'hard work'. One of the 'fathers' of research on unconscious learning, Arthur Reber of Brooklyn College in New York, described in a recent overview of the field how it was that he first came to be interested in it.

> I was drawn to the problem of implicit learning simply because that has always been, for me, the most natural way to get a grip on a complex problem. I just never felt comfortable with the overt, sequential struggles that characterised so much of standard learning . . . As a result of this stance I was not a particularly good 'standard' student . . . I found that what seemed for me to be the most satisfactory of 'learnings' were those that took place through what we used to call 'osmosis', that is, one simply steeped oneself in the material, often in an uncontrolled fashion, and *allowed understanding to emerge magically over time.* The kind of knowledge that seemed to result was often *not easily articulated*; and most interesting, *the process itself seemed to occur in the absence of the effort* to learn what was in fact learned.[13] (Emphasis added)

The studies by Broadbent and Berry, Lewicki and others have made it very clear what learning by osmosis is, what its value is, and the conditions it needs to operate. It extracts significant patterns, contingencies and relationships that are distributed across a diversity of situations in both time and space. It works through a relaxed yet precise non-verbal attention to the details of these situations, and to the actual effect of one's interventions, without any explicit commentary of justification or judgement, and without deliberately hunting for a conscious, articulate mental grasp. Learning by osmosis echoes the insight of the Japanese proverb: 'Don't learn it; get used to it'. It operates in complicated situations which cannot be clearly analysed or defined, and where the goal is to achieve a measure of practical mastery rather than to pursue explanation. And it takes time, as it gradually extracts the patterns that are latent within a whole diversity of superficially different experiences. This form of basic intelligence, inherited from our animal forebears, remains both active and valuable throughout life – if it is unimpeded. It is the first, and the most fundamental, of the slow ways of knowing. Unfortunately, it is all too easy for it to become neglected and overshadowed by d-mode.

CHAPTER 3

Premature Articulation: How Thinking Gets in the Way of Learning

Our simplest act, our most familiar gesture, could not be performed, the least of our powers might become an obstacle to us, if we had to bring it before the mind and know it thoroughly in order to exercise it. Achilles cannot win over the tortoise if he meditates on space and time.

Paul Valéry

About ten years ago, when I was involved in helping people learn to become teachers, I remember sitting at the back of a school science laboratory observing one of my students taking a lesson on photosynthesis. The class of twelve-year-olds had been set a little practical to do, and the student teacher was walking around the lab responding to the pupils' queries. All was going well. Sitting in front of me was a pair of girls working together who had got 'stuck'. They were chatting quietly while one of them kept her hand in the air, waiting patiently for the teacher to notice them and come across to help. The girl who had her hand up was also playing with the fashionable puzzle of the time: the Rubik cube. (This was composed of smaller cubes – each large face having nine such – cunningly engineered in such a way that the faces could be rotated with respect to each other. The mini-cubes were of different colours, and the idea was to manipulate the cube as a whole in such a way that each face of the big cube ended up composed of mini-cubes all of the same colour.)

Having only one free hand, the girl was holding the cube in the other, and turning the faces with her teeth – all the while keeping up her conversation with her friend. She seemed to be giving only

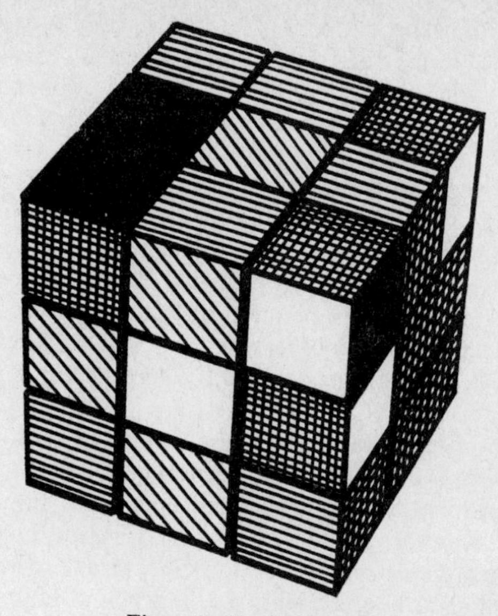

Figure 2. The Rubik cube

the most minimal attention to the manipulation of the cube. Yet, as I watched, I could see that she was making some kind of progress, and every so often stopped to reverse the last few moves and take a different tack. I went over to her and asked her to tell me what she was doing with the cube. She looked startled, both because she thought I might be ticking her off in the indirect way that teachers sometimes adopt, but also because she hadn't realised what she had been doing. It was as if she was surprised to find the Rubik cube in her hand. She looked at me to see if I was 'cross', and on reassuring herself that I was genuinely interested, explained, I think to the best of her ability, what she had been doing. 'Nothing,' she said. 'Just messing about.'

Adults, like myself, were prone to become rather frustrated – even infuriated – with 'the stupid cube', and to feel embarrassed and inadequate in the face of the apparent ease with which children – even not very 'bright' children – seemed able to master it.[1] We could not *understand* how to do it, and after toying with it for a while, we would give it back to its small owner, as if it were too trivial to be worth bothering with, and find something with which to repair the small dent to our self-esteem. The trouble was that we grown-ups went immediately into d-mode, trying to *figure it out*, and, in the case of the Rubik cube, this was not the right mode to be in. It is just too complicated for that. As with Lewicki's patterns, or the incomprehensible sentences, the powers of logic and memory needed were beyond our normal range. What was required, if one was to master the cube, was a gradual build-up of the ability to *see* various recurrent patterns, and to adjust one's moves accordingly: to sharpen our wits through the non-intellectual process of observing and experimenting that we have just discussed. And this is just the kind of 'knowing' that my twelve-year-old scientist's 'messing about' was good at delivering. She had not yet lost the knack of this casual, apparently incidental, way of learning; nor did she seem to mind if she could not articulate its results. I, a long-time d-mode addict, had, and did.

What is the relationship between implicit know-how – the practical intelligence that enables us to function well in the world – on the one hand, and the explicit, articulated understanding that d-mode delivers on the other? It is widely assumed, in education and elsewhere, that conscious comprehension – being able to articulate and explain – is of universal benefit. To understand how and why to do something ought to help us to do it. But does it? In the case of the adults' response to the Rubik cube, it seems as if there is an acquired *need* to understand which may actually block the use of our non-intellectual ways of knowing. We have forgotten them, or do not 'believe' in them any more. There is now good evidence that this suspicion is well founded.

The 'stupid cube' effect appeared in Broadbent and Berry's studies. Not only does people's intuitive ability to control the factory output develop much faster than their ability to explain what they were doing; their *confidence* in their ability tends to follow their explicit knowledge, rather than their know-how. Unless they are able to explain what they are doing, they tend to underestimate quite severely how well they are doing it. People feel as if they are merely guessing, even when they are in fact doing well, and, if they had felt free to, many of the subjects would probably have withdrawn from the game, for fear of looking foolish. It is only because they would have felt even more foolish dropping out that they persevered with the task, despite their lack of confidence – and actually gave their unconscious learning a chance to reveal itself. The subjects have learnt to put their faith in d-mode as the indicator of how much they know, and therefore to distrust, at least initially, perfectly effective knowledge that has not (yet) crystallised into a conscious explanation.

You would at least imagine that there would be a positive link between the two kinds of knowledge, implicit and explicit: that people's sense of having a conscious handle on what they are up to should correlate with how well they are in fact doing. After all, we expect airline pilots and medical students to take written examinations, as well as practical ones, so we must assume that the verbal tests of knowledge and understanding are assessing something relevant. Unfortunately, this does not always seem to be the case. In several investigations of the Broadbent and Berry type, people's ability to articulate the rules which they think are underlying their decisions turns out to be *negatively* related to their actual competence.[2] People who are better at controlling the situation are actually worse at talking about what they are doing. And conversely, in some situations it appears that the more you think you know what you are doing, the less well you are in fact doing. You can either be a pundit or you can be a practitioner, it appears; not always can you become both by the same means.

The situations where this dislocation between expertise and explanation appears most strongly are those that are novel, complicated and to some extent counterintuitive; where the relevant patterns you need to discover are different from what 'common sense' – the 'reasonable assumptions' on which d-mode rests – might predict.[3] In situations where a small number of factors interact in a predictable fashion, and where these interactions are in line with what seems 'plausible' or 'obvious', then d-mode does the job, and trying to figure out what is going on can successfully short-circuit the more protracted business of 'messing about'. But where these conditions are not met, then d-mode gets in the way. It is not the right tool for the job, and if d-mode is persistently misused, the job cannot be successfully completed. Trying to force the situation to fit your expectations, even when they are demonstrably wrong, allows you to continue to operate in d-mode – but prevents you from solving the problem.

For example, consider a classic experiment performed by Peter Wason of the University of London. Undergraduate students were shown the three numbers 2, 4 and 6 and told that these conformed to a rule that Wason had in mind.[4] The students' job was to generate other trios of numbers – in response to which Wason would say whether they did or did not conform to the rule – until they thought they knew what the rule was, at which point they should announce it. Typically the conversation would go like this.

Student: 3, 5, 7?
Wason: Yes (that meets the rule).
S: 10, 12, 14?
W: Yes.
S: 97, 99, 101?
W: Yes.
S: OK, the rule is obviously n, n+2, n+4.
W: No it isn't.
S: (very disconcerted) Oh. But it must be!

The problem is that the students thought the rule was obvious from the start, and were making up numbers only with the intention of confirming what they thought they already knew. If their assumption had been correct, their way of tackling the problem would have been logical and economical. But when what is plausible is not what is actually there, those operating in this manner are in for a nasty shock. In fact, Wason's rule is much more general: it is 'any three ascending numbers'. So '2, 4, 183' would have been a much more informative combination to try – even though, to someone who thinks they know the rule, it looks 'silly'.

When d-mode is disconcerted like this, it often responds by trying all the harder. Instead of flipping into a more playful or lateral mode, in which silly suggestions may reveal some interesting information, people start to devise more and more baroque solutions. 'Ah ha,' they may think, 'maybe the rule is the middle number has to be halfway between the first and the third. So let's try 2, 5, 8 and 10, 15, 20.' When Wason agrees that these too conform to the rule, they heave a sigh of relief – only to be flummoxed once again when they announce the rule and are told it is incorrect. Or, even more ingeniously, they may cling to the original hypothesis – which they have clearly been told is not the solution – by rephrasing it. So they might say, 'OK, it's not n, n+2, n+4, but perhaps it is take one number, add four to it to make the third number, and then add the first and third together and divide by two to get the middle number' – which is, of course, the same thing. Having articulated a misleading account, people then proceed to use this faulty map to guide their further interactions with the task, rather than relying on the ability of trial and error, 'messing about', to deliver the knowledge they need. Attention gets diverted from *watching how the system actually behaves* to *trying to figure out what is going on, and using these putative explanations as the basis for action.*

What happens when you introduce into the Broadbent and Berry task some instruction, in the form of potentially helpful hints and suggestions? Does this give learners a head start, or does it handicap them? Again, conventional educational 'wisdom' would strongly back the former, and once more the research shows that things are not so straightforward. Conscious information is not always an asset, especially when it is given early on in the learning process, or when it serves to direct attention to features of the situation which may be true, but which are not strictly relevant to the way it behaves, or which interact in unexpected ways with other features.

For example, if, in the factory task, I give you a hint that the *workers' age* is worth paying attention to, this information may send you off on a mental wild-goose chase if it eventually turns out that what matters (in this hypothetical factory) is doing the job not too fast and not too slow – and that work-rate is related to age, so that people in their thirties and forties are to be preferred to those in either their twenties (who are too quick) or their fifties (who are too slow). If this correlation is something that would never have occurred to you, then my suggestion has flipped you into what is a doomed attempt to try to understand how age is relevant, and, by the same token, diverted you away from just seeing what happens. People tend to assume, quite reasonably, that the information they have been given ought to be useful, so they keep trying to use it, even when that is not the best thing to do. And in doing so, they may effectively starve the unconscious brain-mind of the rich perceptual data on which its efficacy depends. The time when some instruction *may* be of practical benefit, it turns out, is later on, after the learner has had time to build up a solid body of first-hand experience to which the explicit information can be related.

The fact that giving instruction and advice, in the context of developing practical expertise, is a delicate business, is well known (or should be) to sports coaches, music teachers, and trainers of management or other vocational skills. Most coaches and trainers understand very well that the major learning vehicles, in their lines of work, are observation and practice, and that hints, tips and explanations need to be introduced into learners' minds slowly and appropriately. Whatever is offered needs to be capable of being bound by learners into their gradually developing practical mastery. It must be tested against existing experience and incorporated into it, and this takes time. Coaching is, to draw on my earlier analogy, like making mayonnaise: you need to add advice, like oil, very sparingly. If you add too much, too quickly – if you are in a hurry – then the mind curdles, conceptual knowledge separates out from working knowledge, and you will be on the way to producing (or becoming) a pundit rather than a practitioner.

The corollary of these results is that, when people find themselves in situations where learning by osmosis is what is called for, then they ought to learn better if they have given up trying to make conscious sense of it. If you have abandoned d-mode, it cannot get in the way. A recent experiment by Mark Coulson of the University of Middlesex suggests that this may well be so.[5] He employed two variants of the 'factory' task, in one of which the relationship between the subjects' responses and the system's behaviour was fairly 'logical', and in the other of which it was not. In this second 'illogical' version, the system was programmed to respond in a way that depended on what the subject's response had been one or two trials *previously*, rather than on the current trial – a relationship that does not make a great deal of intuitive sense. (This is somewhat analogous to the party game in which one person has to try to discover, by asking Yes/No questions, the nature of a 'dream' that everyone else has apparently agreed upon. Unbeknownst to the victim, the others respond Yes or No to her questions purely on the basis of whether her question ends in a vowel or a consonant. The fun comes from the fact that this rule starts to generate some fairly bizarre information about the 'dream', and that the more the victim tries to make rational sense of this information, the stranger the 'dream' becomes, and the less likely she is to discover the 'trick'.)

Similar studies have shown that the logical task is amenable to the d-mode approach, while the illogical task is not. As with the dream game, the correlations between question and answer in the illogical version are so obscure that the attempt to follow sensible lines of thought and construct reasonable hypotheses about what is going on is unlikely to uncover them. The only effective strategy is to try to observe what is happening with as few preconceptions as possible. Thus subjects should do better on the illogical task if they have somehow been persuaded to give up d-mode before they start. Conversely, if they have abandoned d-mode they should do worse on the logical version.

Subjects in Coulson's study took part in either the 'logical' or the 'illogical' version of the task. Their job, as before, was to learn, over a series of trials, to control the factory process. However, in each version, half of the subjects had been given some advance 'training' in which the behaviour of the computer was *completely random* – an experience which, Coulson reasoned, would have weakened their faith in d-mode, as no amount of clever thinking could reveal patterns where there were none to be found. While overall subjects took longer to learn the illogical than the logical version, the group who had had the prior 'random' experience learnt the illogical task faster than those who had not. Subjects who had had the random experience, on the other hand, learnt the logical version more slowly than those who had not. Coulson argued that the preliminary experience of grappling with the random version of the task induces a state of confusion, so that, when the main task comes along, subjects have dropped d-mode in favour of a learning by osmosis approach. If the main task is actually the illogical one, this puts you at an advantage. Your learning-by-osmosis is unimpeded by the intellect. But if the main task is one which is amenable to being figured out, then you are disadvantaged if you have abandoned d-mode. Whether to back the hare or the tortoise depends crucially on the nature of the situation. If it is complex, unfamiliar, or behaves unexpectedly, tortoise mind is the better bet. If it is a nice logical puzzle, try the hare brain first.

There are indeed many cases in which d-mode is the right tool, and in which the hare clearly comes out the winner. Imagine that you have a regular chessboard, an 8 × 8 chequered square, and you cut out two diagonally opposite corner squares (leaving 62 squares, see Figure 3). You make up 31 domino-shaped bits of cardboard, each of which neatly covers two squares on the board. You give me the mutilated board and the oblong pieces of card and ask me if I can exactly cover the 62 squares using the 31 bits of board, without cutting, bending or overlapping them. What do I do?

My first thought may be that of course I can do it – 31 dominoes, each covering two squares; 31 × 2 = 62; QED. Your quizzical look, however, strongly suggests that it is not quite that simple. So what I then do is start laying out the dominoes on the board . . . but every time I try it, I always seem to be left with an odd square on the opposite side of the board from the available piece of cardboard. As I am deep down convinced that it *is* possible, I keep shuffling the dominoes around hopefully; but finally have to confess that I do not seem to be able to find a solution. A large amount of time, and some emotional energy, are consumed.

You then invite me to think about the *colours* of the squares . . . especially the ones that have been cut out. I, having implicitly decided that the colours are irrelevant to the problem, and therefore given them no thought, wonder what you are talking about. Then I realise that the two opposite corner squares must be the *same* colour, either both black or both white. If you have taken two white

Figure 3. The mutilated chessboard

squares away, that means that there are 30 white ones and 32 black ones left – an unequal number. But each domino has to cover two adjacent squares, i.e. one black one and one white one. So for the puzzle to be soluble there has to be not just an even number of squares, but an equal number of blacks and whites. Obviously – now I come to think about it – it can't be done. (Imagine a 2 × 2 board consisting of four squares; take away diagonally opposite squares and, by analogy, the answer is plain.) Some straightforward deliberation could have saved me time and trouble.

Pawel Lewicki's group at the University of Tulsa have investigated a slightly different aspect of the relationship between know-how and knowledge: whether they automatically change together, or whether learning that affects one can leave the other unchanged. The researchers focused on one particular set of patterns that we have all been developing since babyhood, and in which we might be considered to be quite expert: those that associate how people look with how they are likely to react – their faces, most obviously, with their moods and personalities. Even if much of this knowledge is implicit, we should have developed some conscious self-knowledge about the interpersonal rules of thumb that we tend to use. Spectacle wearers are likely to be studious, for example. People who don't make eye contact are shy or shifty. People whose eyes have large pupils are more warm and friendly than those with small pupils. People whose heads loll about in an alarming way are probably not Cambridge professors. Everyone has their idiosyncratic set of diagnostic features. We think we can recognise 'sad eyes'. 'mean mouths' or 'business-like moustaches'.

Lewicki first elicited from his subjects as many of these personal associations as they could give. He then asked them to look at a long succession of photographs of unfamiliar people, and to try to predict what their personalities were like. After each picture, they were given 'feedback' about how good their predictions were. Unbeknownst to the subjects, Lewicki had again 'stacked the deck', by determining the character which he attributed to each photographed person on the basis of some subtle combination of facial features. As with the experiments of his which were discussed in Chapter 2, subjects gradually got significantly better at making the predictions, even though they had absolutely no conscious knowledge of any connection between the facial features and the supposed personalities.

However, there is a new twist. Lewicki had rigged his character attributions so that, for each subject, some of the connections between face and personality were the exact opposite of the ones which they had told him they relied on in everyday life. So in order to learn the patterns in the experiment, they were required to go against their normal assumptions. What effect did this have, Lewicki asked, either on the speed with which the subjects learnt the experimental pattern, or on the strength of their pre-existing rules of thumb? Should the mismatch not slow down the new learning, and/or cause some shifting in what is known consciously? You would think so – if you make the commonsense assumption that people's self-knowledge is an accurate reflection of the way they go about things.

In fact, Lewicki found that the subjects' pre-existing conscious beliefs a) had no effect on the speed or efficiency with which the contrary associations were learnt through experience; and b) were themselves unaffected by the unconscious learning that had taken place. The undermind is acquiring knowledge of which consciousness is unaware, and by which it is unchanged, and using it to influence the way people behave. Consequently a schism develops between what people think they know (about themselves), and the information that is unconsciously driving their perceptions and reactions. The views that they *espouse* about themselves, we might say, become at odds with the ones that their behaviour in fact *embodies*.

This small experiment thus furnishes us with a neat illustration of the kind of 'split personality' phenomenon with which we are all familiar, but which it is often convenient to ignore: the existence in the mind of a second centre of operations which is capable of going its own way, untroubled by what conscious 'headquarters' happens to be saying. And consciousness itself can remain unruffled by the discrepancy, by the simple expedient of not noticing that it exists. At the end of a review of his experiments, Pawel Lewicki concludes that 'our non-conscious information processing system appears to be faster and "smarter" overall than our ability to think and identify meanings . . . in a consciously controlled manner. Most of the "real work" [of the mind] is being done at a level to which our consciousness has no access.'[6] This, from the hard-nosed world of cognitive science, is an extraordinary conclusion. Yet it is what these carefully controlled experiments reveal.

The studies we have reviewed so far demonstrate that the urge to be articulate is a mixed blessing when it comes to learning. But there are other areas of life where the same might be said. What about the execution of a skill that is already well learnt, for example? Does it make any difference to one's expertise whether one is able to put what one knows into words, or not? In a paper entitled 'Knowledge, knerves and know-how', R. S. Masters of the University of York has shown that people who can articulate what they are doing may go to pieces under pressure more than those whose skill in entirely intuitive.[7] He studied people who were learning to play golf, focusing particularly on their putting skill. One group of learners was taught how to putt 'explicitly' – they were given a set of very specific instructions which they were asked to follow as carefully as they could. Another group was given no instructions – they simply practised – and they were even asked to occupy their minds with an irrelevant task to prevent them thinking about their putting as they were doing it. After their training, both groups were tested on their putting ability by an imposing 'golfing expert' whom they had not met before; and there were also significant financial rewards and penalties depending on how well they did. Both the 'expert' and the money were designed to make the test stressful.

Masters discovered that the performance of those who had learnt intuitively held up much better than that of those who had been following instructions. His explanation was that the breakdown of performance under pressure – what sportspeople refer to as 'choking', or 'the yips' – was due to the instructed group flipping back into d-mode and trying to remember and follow instructions, rather than just play the shot. People who had learnt intuitively were not able to do this, as they had no explicit knowledge to fall back on. They just had to carry on as normal – and this, it turned out, stood them in better stead. Thinking about what you are doing may introduce a kind of analytic self-consciousness which gets in the way of fluent performance – an effect reminiscent of the famous centipede who was rooted to the spot when asked which leg he moved first.

Know-how is acquired in different ways from verbal knowledge, as we have seen; but it is also 'formatted' differently, and is good for different kinds of purposes. For example, Euclidean geometry – the kind we did at school – is an extremely elegant and powerful tool for describing a family of idealised shapes, those that can be made out of straight lines and mathematical curves drawn on a flat surface, and regular three-dimensional objects such as spheres, cubes and cones. In this arcane universe, all kinds of strange and beautiful properties appear, and precise calculations can be made. The areas of circles and parallelograms, for example, can be computed exactly with the aid of certain formulae. However, if you ask geometry about the area of an *untidy* shape, one that cannot be described by equations, it immediately loses its power and grace. The real, irregular world is too awkward and intractable, and it has to be neatened up, in the way the axioms of geometry demand, before it can be treated. To calculate the area of France, using Euclid, we would have to suppose it to be a badly drawn hexagon, or to superimpose upon it a grid of little squares. Unless we force it into such *a priori* shapes and categories, we cannot get our strong generalisations to work.

Now, in contrast, consider a humble device called the polar planimeter, invented in 1854 by a German mechanic, Jacob Amsler.[8] It consists of two sticks flexibly joined together as in Figure 4. The top end of the 'vertical' stick is fixed to the table. At the left-hand end of the 'horizontal' stick is a wheel that sits on the table which can both rotate and skid (if it is pulled sideways). At the right-hand end is a pointer that also sits on the table. The cunning thing about this simple tool is that, if you trace the outline of any shape with the pointer, the wheel will rotate by an amount that is directly proportional to the area of the shape. All you have to do is calibrate

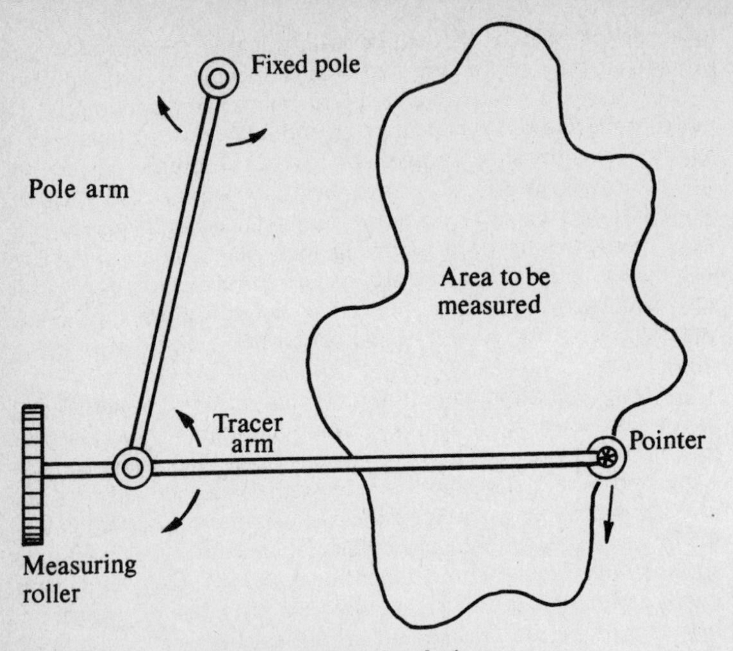

Figure 4. The polar planimeter

the wheel by tracing out a shape with a known area – a 5cm square, say – and then you can use your polar planimeter to measure the area of any shape, no matter how odd.

The knowledge which d-mode generates, and on which it relies, is more like Euclidean geometry. It tends to be general, abstract and powerful, and to apply it to particular cases you often have to make the world appear neater than it truly is. The polar planimeter corresponds to 'know-how'. It capitalises intelligently on a curious fact, and uses a 'trick' to solve cheaply and easily what for geometry is embarrassingly difficult. Geometry can do lots of things that the polar planimeter cannot, but for the particular job of measuring irregular areas (within a certain range of sizes) the planimeter is much more accurate and efficient. *How* it does it neither I nor Herr Amsler can tell you – and provided it works, our explanatory shortcomings are of little consequence (though they pose a nice intellectual challenge: the basis for a PhD perhaps). It is said that the painter Giotto could draw perfect freehand circles, and would leave them as calling cards. It is doubtful whether he knew the algebraic formula for a circle, or how to calculate its circumference, and certain that a course in geometry would have not improved his skill, and might well have impeded it.

Our know-how is in general of this ad hoc, smart, opportunistic kind. The know-how regions of our minds are organised less like the Library of Congress than a well-used kitchen: logic continually gives way to convenience. I do not have to make my kitchen so rational that anyone could come in and figure out where the Tabasco sauce is from first principles. If I did have such a logical layout, I would not be so well set up to cook the kinds of things that *I* like to cook, and habitually do. Know-how is, as I say, formatted differently to knowledge in that it grows by osmosis (rather than comprehension); manifests itself in specific domains of expertise (rather than in abstractions); capitalises on serendipity (rather than first principles); and is organised idiosyncratically (rather than systematically). No wonder that the ways of knowing that use and create it are very different in their time characteristics from d-mode.

Western culture's over-reliance on d-mode reflects a lack of appreciation of these vital differences between knowledge and know-how. We tend, as a society, to make what was originally only an academic error: what Pierre Bourdieu refers to as the *scholastic fallacy*. 'This fallacy . . . induces people to think that agents involved in action, in practice, in life, think, know and see as someone who has the leisure to think thinks, knows and sees.'[9] By assuming that knowledge is similar to know-how, we are led to suppose that know-how can – even should – be acquired through knowledge, and that knowledge, once acquired, ought to transform itself automatically into know-how. Managers are sent on a five-day course on 'leadership', and are immediately supposed to come into work the following Monday and start leading. The frustration with, and frequent cynicism about, such short courses, in the business world and elsewhere, is not due to lack of commitment on the part of participants, nor of skill on the part of the trainer. It reflects a deep confusion about the nature of learning and knowing.

The confusion makes us promote 'book learning' and formal education (and training) as the proper medium for acquiring everything. Adults pore over the instruction manual for a new computer, afraid to plug it in until they know how it works and what to do, while their children have already discovered, just by 'messing about', how to make it do the most complicated tricks. Apprentice midwives used to learn their craft by assisting their more experienced mentors at hundreds of births. Now they have to have a degree. There are even those who argue that couples should have to attend a series of seminars on 'parenting skills' before they are allowed to have a baby. The tragedy is that now there may even be some sense to this. If the other ways of knowing *have* been effectively disabled by the belief that intellect is the only mode we have, or the only mode we need, then the belief becomes the reality. D-mode does then provide the only avenue open to us for learning, however limited or inappropriate it may actually be for the job in hand.

A MORI poll in 1996 on learning attitudes revealed that two-thirds of people 'prefer to learn from books', while another 19 per cent prefer CD-ROM and computers. Nobody, apparently, said that they prefer to learn by messing about, by osmosis, or just by watching. Learning has become something that you do in a special place, with special equipment, under the instruction of experts, using your deliberate, conscious intellect. No other possibilities seem to be catered for: a pity, at the very least, if learning by osmosis is, when faced with certain kinds of complexity, a more intelligent option than d-mode.

But learning by osmosis has its own limitations, just as d-mode has. Not only may it be deployed at the wrong time, leading to a protracted process of trial and error which could have been short-circuited by a little logical thought; it often cannot be communicated, or only very crudely, and there are many occasions on which this is a definite handicap. The first time I went ice-skating with my twelve-year-old cousin, I strapped on the skates and stood nervously on the edge of the rink sliding my feet backwards and forwards, convinced of the physical impossibility of what dozens of people were doing around me. Eventually I swallowed my pride and asked Dany to tell me how it was done. 'It's easy,' she said, 'watch,' and she sped off round the ice. When she got back I was beginning to get irritated. 'I know you can *do* it,' said, 'but I want to know *how* to do it.' 'It's easy,' she said, 'watch,' and sped off again.

Her know-how was completely inarticulate, yet there *are* useful tips and explanations that can be given. To be a practitioner, it may be best not to be able to think too much about your skill or your art. But to be a coach is different. A whole new phase of learning may be required if the virtuoso wants to become a teacher, for she may have laboriously to unpick her seamless expertise and turn it into the descriptions and explanations that, judiciously administered, help learning to happen.

The most important limitation of know-how, however, is its relative inflexibility. Practical knowledge that has been learnt without thinking may work smoothly and fluidly within the original domain.

But many psychological studies have shown that when the appearance of a task is changed, even if only a little, while the underlying logic remains exactly the same, know-how often fails to transfer. People who have learnt to control the factory process may function no better than a complete beginner if effectively the same problem is now presented as being about the control of 'traffic flow'. The polar planimeter is useless for determining the volume of three-dimensional objects, while the principles of Euclidean geometry can easily be extended. With know-how, perceiving and doing are wrapped up together in one tightly interwoven package.

From an evolutionary perspective, the 'bundling' of know-how does not matter at all if your world consists of a small number of separate scenarios – looking for food, cleaning your fur, mating, sleeping, avoiding predators and raising your young, for example. If your life consists of such a neat set of discrete jobs, then your main problem, apart from keeping your wits sharpened, is to know which scenario you are in, or which one you need to switch to. To have your know-how organised and integrated under separate headings is economical and efficient. But if your world is more complicated, the scenarios or 'scripts' you take part in become more numerous, and they begin to interweave. The same individuals in your community may take different roles in different scripts. For the male black widow spider, or praying mantis, your mate may suddenly turn into your executioner.

As life gets more intricate, so it becomes a matter of survival to be able to deconstruct situations into familiar parts, and to be able to construct responses to hybrid situations by putting together different facets of different scripts. Parties, for example, may be stressful because they bring together friends and family with whom one has quite different kinds of relationship, and which separately bring out contrasting sides of your personality. If you could only relate to them 'in context', and had only a single stereotyped 'party' script, you would have no way of solving this intricate social problem. But if you have a sense of your friends that is somewhat disembedded from the contexts in which you usually meet, you may be able to integrate all the pieces that originally came from different 'jigsaw puzzles' into a new and hopefully coherent picture.

This carving of scenarios into recombinable 'concepts' is, basically, the ability conferred by language, and by d-mode. When understanding has been 'articulated', it has not only been turned into words. Articulated also means 'jointed; composed of distinct parts which may move independently of each other'. Know-how is not articulated in either sense. It is not capable of being taken apart, reflected upon, or put together again in novel ways when expertise breaks down or conditions change. It can only shift gradually under the influence of learning by osmosis. And because it cannot be discussed, it cannot easily be influenced by what other people may say or bring to it. The risk with fluent know-how is that it will be deployed *mindlessly*, in a way that takes no account of considerations or information that is held in a different sub-compartment of the mind. The ability to see that some aspect of what you have learnt in one situation is of relevance in another which looks different superficially is a highly valuable one, and it has been shown in several experiments that it can be increased by the use of conscious reflection.

In one classic study of what has come to be referred to as 'functional fixedness', people were set a problem which could only be solved by seeing that a familiar object could be used in an unfamiliar way. The task was to tie together two pieces of rope hanging from the ceiling. The problem was that they were too far apart to be grasped simultaneously. In the experimental room there was a variety of everyday objects, including a pair of pliers. The problem could be solved by seeing that the pliers could be used as a pendulum bob: you tie them to the bottom of one of the ropes and set it swinging so that, when you are holding the other rope, the first now comes within your grasp. Left to themselves, a high proportion of people fail to solve the problem. But if the experimenter waits until they have got stuck and then simply says 'Think! Think!', many subjects then spontaneously see the solution.

Without d-mode, without the benefits that concentrated analytical awareness bring, the lower animals are smart, but within limits. The spider, *sphex*, the digger wasp, and even the gobiid fish are good at what they are designed by evolution to do: they meet a range of challenges with conspicuous intelligence, but when the world throws a different *kind* of challenge at them, they are found wanting: inflexible and uncreative. They cannot turn around what they 'know' and recombine it in ways that are both novel and appropriate. They cannot dismember their abilities and perceptions – cannot segment and articulate them – and so cannot *re*-member them to suit an unprecedented present. Their unconscious intelligence is more or less crystallised; they lack the ability to dissolve it and reconstitute it.

The same is true, initially, of children, but they are able to transcend these limitations. As they develop, the range and complexity of the scenarios in which children take part start to expand dramatically. They go to playgroup and on to school, where they meet different kinds of adults with different agendas, and with whom they have very different kinds of relationship from those they have with their parents. They take part in new social groups of various sizes and compositions. They start to meet many different *kinds* of things to be learnt about, and to discover new ways of going about learning them. And as they do so, they face a choice: whether to keep multiplying the number of separate mental scenarios; or whether to start to seek a higher-order level of knowing that enables these different scripts to be integrated, compared and combined. If they take the former option, their mental landscape develops into a patchwork of separate 'modules' of know-how that are unable to share what they know among themselves. If they take the latter they need to develop a new form of learning, one which enables them to *ruminate* over their experience; to bring back, as the cow does, what has been separately ingested, and by chewing it over make it more homogeneous. They would have to be concerned not just to meet new challenges one by one, but to look actively for points of segmentation and integration.

And children do start to develop this ability to ruminate, it turns out, around the age they first go to school. Annette Karmiloff-Smith of the Medical Research Council Cognitive Development Unit in London has demonstrated the beginnings of rumination in the context of what she calls 'learning beyond success'. Across many different kinds of task, she has observed that children will first learn how to 'get it right' – and will then, if they are given the opportunity, continue to 'play' with the situation in ways that actually reduce their apparent control and competence for a while. In language learning, for example, a child will very often learn to say (correctly) 'went', but will then go through a phase of using the 'regular' (but wrong) form 'goed', before finally reverting to 'went'. Or they will learn to balance different-shaped rulers on a fulcrum, and then make mistakes, and then get it right again.

Karmiloff-Smith argues that these dips in performance are symptomatic of exactly the kind of searching for coherence and conceptualisation that I have described. It is as if, when faced with a challenge, children use whatever is at hand to respond to it, like someone after a shipwreck constructing an emergency raft out of all kinds of flotsam in order to keep afloat. But later, when they have a little more leisure, after the storm has passed, they move into a more reflective mode in which they experiment with taking this lash-up to bits again to see what happens, and where it might fruitfully draw on pre-existing pockets of know-how developed to cope with different situations, to make their know-how as a whole more elegant, integrated and powerful.

Language, and the ways of knowing which it affords, liberates; but it comes with snares of its own. Although it allows us to learn from the experience of others, and to segment and recombine our own knowledge in novel ways, it creates a different kind of rigidity. As Aldous Huxley said: 'Every individual is at once the beneficiary and the victim of the linguistic tradition into which he has been born – the beneficiary inasmuch as language gives access to the accumulated records of other people's experience; the victim insofar as it . . . bedevils his sense of reality, so that he is all too apt to take his concepts for data, his words for actual things.'[10] Know-how is tied to particular domains and purposes, but within those bounds it is detailed, accurate, efficient and flexible. D-mode creates a superordinate stratum of knowledge that transcends particular contexts, but is, by the same token, more abstract, and liable to become detached from the shifting layers of experience that originally underpinned it. As the Lewicki experiment showed, once this detachment has taken place, know-how can develop pliably in response to new exigencies of experience while knowledge is left unaltered, cast in stone.

Language is not only the internal code in which knowledge is inscribed; it also relies upon, and enshrines, a public system of categories. A language represents a consensus about how the world is to be segmented, and thus determines heavily how things are categorised, talked about, and even perceived. Much has been written about the relationship between language and 'reality', but the only point to note here is that we are obliged to articulate our know-how in terms that we ourselves have not chosen, and which may well not be the most congenial or accurate descriptors of our personal experience. As we articulate our experience, so we have to pour what is intrinsically fluid and ill-defined into moulds that are more clear-cut, and not of our own making. The language of d-mode implies a 'reality' that is neater, more solid, more impersonal and more agreed-upon than the one that often confronts us. It is both an approximation – leaving out much of the detail – and a distortion – introducing fictional elements that actually have no referent.

D-mode is like map-reading: with a map we are able to get our bearings, and see how one area relates geographically to another.

But maps must be simpler and more static than the world they represent; and they contain conventions that aid the interpretation of the map, but which are not 'real'. As we climb the mountain, we do not periodically have to step over the contour lines. As we cross from England to Wales, the terrain does not change from pink to blue. It is not the case that we cannot go where there is no track, nor, certainly, that the motorway is always the best route. When the map is good enough, and we understand the status of the conventions, then d-mode works well. When we forget, as Alfred Korzybski[11] insisted, that 'the map is not the territory', or when we need, to resolve a predicament, a finer-grain, more subtle or more holistic image than language provides – it is then that we need recourse to our other, slower ways of knowing. Some predicaments cannot be dealt with effectively with the tools of analysis and reason. And there are some, too, that will not succumb to an increase in expertise, such as learning by osmosis delivers. To deal with such problems, we need access to those slow ways of knowing we have preliminarily called rumination or contemplation; mental modes which deliver, it is claimed, forms of *creativity* and *intuition*.

CHAPTER 4

Knowing More than We Think: Intuition and Creativity

'Did you make that song up?'
'Well, I sort of made it up,' said Pooh. 'It isn't Brain . . . but it comes to me sometimes.'
'Ah,' said Rabbit, who never let things come to him, but always went and fetched them.

A. A. Milne, The House at Pooh Corner

In his autobiography the nineteenth-century English philosopher Herbert Spencer recounts a conversation with his friend Mary Ann Evans – the novelist George Eliot. They had been discussing Spencer's recently published book *Social Statics*, and George Eliot suddenly observed that, given the amount of thinking he must have done, his forehead remained remarkably unlined. 'I suppose it is because I am never puzzled,' said Spencer – to which Eliot, understandably, replied that this was the most arrogant remark she had ever heard. Spencer says that he went on to justify his remark by explaining that

> my mode of thinking does not involve the concentrated effort which is commonly accompanied by wrinkling of the brows. The conclusions, at which I have from time to time arrived . . . have been arrived at unawares – each as the ultimate outcome of a body of thoughts that slowly grew from a germ . . . Little by little, in unobtrusive ways, without conscious intention or appreciable effort, there would grow up a coherent and organised theory. Habitually the process was one of slow unforced development, often extending over years; and it was, I believe, because the thinking done went on in this gradual,

almost spontaneous way, without strain, that there was an absence of those lines of thought which Miss Evans remarked.

In Spencer's opinion 'a solution reached in the way described is more likely to be true than one reached in pursuance of a determined effort to find a solution. *The determined effort causes perversion of thought*. . . An effort to arrive forthwith at some answer to a problem acts as a distorting factor in consciousness and causes error, [whereas] a quiet contemplation of the problem from time to time allows those proclivities of thought which have probably been caused unawares by experience, to make themselves felt, and to guide the mind to the right conclusion.' (Emphasis added)[1]

The ways of knowing with which both Pooh Bear and Herbert Spencer are familiar are different from d-mode in a number of ways. Most obviously, they take time, and therefore they require patience: a relaxed, unhurried, unanxious approach to problems. In this they resemble 'learning by osmosis', but they are not the same. In learning by osmosis, the undermind gradually uncovers patterns that are embedded in, or distributed across, a wide variety of experiences. Know-how is distilled from the residue of hundreds of specific instances and events. But while Spencer's insights into the organisation of society undoubtedly drew on much prior thought and observation, the process that he is referring to is one that goes beyond this unconscious distillation. This process seems to reflect not the acquisition of new information so much as the mind's ability to discover, over time, new patterns or meanings within the information which it already possesses, and to register these consciously as *insight* or *intuition*. Though experience provides the data, the process is not acquisitive but ruminative. Pooh's song that prompted Rabbit's question demonstrates the same process on a smaller scale. He was not announcing the inductive discovery of a new generalisation, but simply producing something which came 'out of the blue'.

Despite the widely-held assumption that d-mode represents the most powerful thinking tool we possess – which makes it the one we call upon, or revert to, in the face of urgent demands for solutions – the truth is that our ideas, and often our best, most ingenious ideas, do not arrive as the result of faultless chains of reasoning. They 'occur to us'. They 'pop into our heads'. They come out of the blue. When we are relaxed we operate very largely by intuition. We don't usually offer a detailed rationale for our restaurant preference: we say 'I feel like Thai'. We happily allow ourselves to be nudged by feelings and impulses that do not come with an explicit justification. Yet when we are put 'on the spot' in a meeting, or are faced with an urgent 'problem' that demands 'solution', we may act as if these promptings were weak, unreliable or negligible. We feel as if intuition will not stand up to scrutiny, and will not bear much weight. There is now a body of research which shows that intuition is more valuable and more trustworthy than we think; and that we disdain it, when we are 'on duty', to our practical detriment.

We need a more accurate understanding of the nature and status of intuition: one which neither under nor overvalues it. Those who disparage intuition are reacting, often unwittingly, against the presumption that intuition constitutes a form of knowledge that is 'higher' than mere reason, or even infallible. The dictionary definitions still carry some of that inflated view, and by doing so they create expectations that are patently false. Chambers' dictionary gives intuition as 'the power of the mind by which it *immediately perceives the truth* of things, without reasoning or analysis'. The Shorter Oxford is more poetic and more presumptuous still: it gives intuition as 'the immediate knowledge ascribed to angelic and spiritual beings, with whom vision and knowledge are identical'.[2] Now while it may be the case that there is a certain quality of intuition, one which may take much careful cultivation to acquire, which does give access to a qualitatively different kind of knowledge, it is self-evident that everyday intuition falls far short of this ideal. Our promptings are notoriously fallible, whether they concern a career move or a life partner, a book that we misjudged by its cover or a new route that the 'nose' confidently said was a short cut, but which only succeeded in getting us lost.

Intuitions can be wrong, but that does not mean they are worthless. Intuitions are properly seen as 'good guesses'; hunches or hypotheses thrown up by the undermind which deserve serious, but not uncritical, attention. They offer an overall 'take' on a situation that manifests not – not yet – as a reasoned analysis, but as an inkling or an image. Behind the scenes, the undermind may have integrated into this tangible prompt a host of different considerations, including analogies to past experience and aspects of the present situation, of which the conscious mind may not have even been aware. And this integration can happen, as the dictionary definitions say, 'immediately', or it may take time – even, as in Spencer's case, up to years. But the result, when it does 'pop up', is always provisional. It is a pudding, served up by the unconscious, whose proof is in the eating: a critical testing which may be the reaction of the audience to an impromptu witticism, à la Pooh, the rigorous checking of logical implications, or the detailed working out of a creative poetic or artistic theme.

Fast intuitions – 'snap judgements' and quick reactions – are vital responses for the human being, just as they are for animals. When the present event is a variation on a familiar theme, it pays to be able to classify it and react in habitual fashion. To spend time pondering on insignificant details is sometimes wasteful, or even dangerous. No need to inspect the number plate of the bus as it bears down upon you. But these reflexes work to our detriment when a new situation looks similar to ones we have experienced in the past, but is actually different. Then the balance of priorities shifts, and it is now the quick, stereotyped response that is the risky one, while more leisurely scrutiny can pay dividends.

The importance of this shift from fast to slow thinking was graphically demonstrated in the laboratory by Abraham and Edith Luchins as long ago as the 1950s. They set people puzzles of the following sort. 'Imagine that you are standing beside a lake, and that you are given three empty jars of different sizes. The first jar holds 17 pints of water; the second holds 37 pints; and the third holds 6 pints. Your job is to see whether, using these three jars, you can measure out exactly 8 pints.' After some thought (which may, to start with, be quite logical), most people are able to end up with 8 pints in the largest jar. Then they are set another problem of the same type, except this time the jars hold respectively 31, 61 and 4 pints, and the target is to get 22 pints. And then another, with jars holding 10, 39 and 4 pints where the target is 21 pints. (You may like to try to solve these puzzles before consulting the notes for the solutions.)[3] You will find that the same strategy will work for all three problems. But now comes the critical shift. You are next given jars of capacity 23, 49 and 3 pints, and asked to make 20 pints. If you have stopped thinking, and are now applying your new-found rule mindlessly, you will solve the problem – but you will not spot that there is now a much simpler solution. The problem looks the same, but this particular one admits of *two* solutions, one of which is more elegant and economical than the other.

We can easily imagine a business company – or any other kind of organisation – falling into the same trap. They may 'think they are thinking' about each problem as it comes along; but if they are unable to think *freshly*, they will keep coming up with the same kinds of answers – even when circumstances have changed and new possibilities are there to be discovered. And one of the strongest forces that prevents the discovery of these new avenues may be the habit of thinking fast: of taking your first intuitive assessment of the situation for granted, and not bothering to stop and check. Milton Rokeach tested this hypothesis, using the Luchins jars, by forcing people to slow down when they were looking at the new problems. If they were allowed to give the 'solution' in their own time, most people immediately applied the rule that had worked previously without question. But when they were prevented from writing down their answer for a minute, some of them pondered the problem in greater detail – and were able to discover the new solution.

Not surprisingly, this benefit only accrued to people who did actually attend to the details of the new problem during the delay. Many people reported that they made up their minds quickly about the answer, and then spent the enforced interval thinking about all kinds of unrelated things – 'making plans for Saturday night's party', 'thinking about letters I had to write', 'counting the holes in the tiles on the ceiling', and so on – and for them, the extra time obviously did nothing to improve their creativity. What was more interesting, however, was the mental activity of the subjects who did find the new solution. They were *not* earnestly figuring out the answers, or making calculations on bits of scrap paper. They were actually musing in a much more general way on what *type* of questions were being asked, and what the experimenter was up to. One said, 'I was wondering what the experiment was trying to prove.' Another said, 'I was thinking what the results would indicate.' It was this kind of 'meta-level' questioning that led to the insight, not the disciplined application of procedures.

Let me illustrate how intuition works with the aid of a slightly more complicated example (one, incidentally, that Wittgenstein was fond of using in his philosophy seminars). Imagine that the Earth has been smoothed over so that it forms a perfect sphere, and that a piece of (non-elastic) string has been tied snugly round the equator. Now suppose that the string is untied, and another 2 metres added to the total length, which is then spaced out so that the gap which has been created between the string and the Earth's surface is the same all the way round. How big is this gap? Could you slide a hair under the string? A coin? A paperback book? Could you crawl under it? Most people's strong intuition is that the gap would be tiny, of the order of a millimetre or two at the most. In fact, it is easy to prove mathematically that it is about 32 centimetres, or just over a foot; so you could indeed crawl under the string. (The proof is in the notes, for those who wish to follow it.)[4] The strange thing, when you work out the geometry, is that the size of the gap turns out to be independent of the size of the original sphere (or circle: the problem is not essentially three-dimensional). So you would get the same-sized gap whether you started with a tennis ball, a circus ring, or the universe. Most people's intuition, on the other hand, insists that the larger the original object, the less 'difference' the 2 metre extension is going to make: in other words, the smaller the gap.

Intuition goes awry here because it is based on the unconscious assumption that this situation is analogous to other, apparently similar, situations where the idea that 'the larger the object, the smaller the change' *does* apply. If we were to change the puzzle slightly, and say: 'Supposing the oceans were neatened up into a huge cylinder, how much would the level rise if we added 20 litres of water?', then the answer is indeed 'not very much'; and we would be right, in this instance, to assume that the larger the original volume, the smaller the difference to its depth. The 20 litres would make much more difference to the depth of a paddling pool. It just turns out that this plausible assumption works for the height of cylinders, but not for the radius of circles. It is a good guess that in one case turns out to be right, and in the other case wrong. Fast intuitions depend on the undermind taking a quick look at the situation and finding an analogy which seems to offer understanding and prediction. These unconscious analogies surface as intuitions. Whether they are right or not depends not on how 'intuitive' they are, but on the appropriateness of the underlying analogy. Often we are absolutely right. But sometimes the undermind is fooled by appearances, and then it leads us off in the wrong direction.

This example also demonstrates how the way of knowing you employ may give different answers to the very same question. D-mode and intuition may well draw on different processes, knowledge and beliefs, and thus may produce conflicting solutions. If you followed the mathematics in the notes, then you might be rationally persuaded that the gap is a foot, while intuitively you persist in believing it to be minute. Below the surface, some assumptions are being made that lead to one answer. Above the surface, so it seems, different premises lead to a different answer. In this case, it turns out that the 'rational' answer is the right one. In other cases (as when intuition told you that there was something suspicious about the well-spoken woman at the door 'collecting for charity', but you persuaded yourself you were being 'silly') it may be intuition that is right and reason that is wrong. It is an empirical issue.[5]

Which mental mode is engaged – and therefore which answer you get – may depend on how you happen to be thinking when the question arises; or on some – possibly quite incidental – feature of the situation. If you catch a physics undergraduate in the bar one night and ask her why, when you throw a ball, it moves through an arc, she is likely (if she can be bothered) to tell you a story about the 'energy' or 'impetus' you give to the ball when you throw it, and how this gets used up overcoming the drag of the air and the force of gravity. When the upwards 'oomph' has been depleted to a certain level, she says, gravity starts to 'win', the ball reaches its zenith and begins to fall. However, if you then remind her that this is a *physics* problem, she may well stop for a moment (as she switches from intuitive mode to physicists' d-mode) and say, 'Of course. Silly me. There isn't any "oomph" you put in to the ball as you throw it. The only forces are gravity and the air resistance.'[6] Her first 'take' is an everyday, intuitive one; her second switches her into a different frame of reference, giving access to a different database and different ways of thinking. If the question had been on an examination paper, she would have selected d-mode automatically.

The power of context to flip people into one way of knowing rather than another – and to produce quite different responses to what is logically the same problem – is widespread, and very striking. In a study of ten-year-olds by Ceci and Bronfenbrenner in 1985, for example, the children sat in front of a computer screen in the centre of which one of a variety of geometric shapes would periodically appear.[7] Their job was to predict (by moving the cursor with a mouse) in which direction, and how far, the shape was about to 'jump'. The shapes were circles, squares and triangles that could be dark- or light-coloured, and large or small. In theory, the children could have predicted the jump on the basis of the shape, because squares always went to the right, circles to the left, and triangles stayed in the middle; dark things went up and light things down; and large things went a short distance and small things a long distance. After 750 trials, the children had learnt virtually nothing.

However, after making a small change to the task, *which had no effect at all on its logical difficulty*, things looked very different. All the experimenters did was replace the three geometrical shapes with animals (birds, bees and butterflies); swap the normal computer cursor for an image of a 'net'; add some sound effects; and tell the children that this was a game in which they had to try to catch the animals as they moved. After less than half as many goes, all the children were placing the net in the right position to 'capture' the animals with near-perfect accuracy. The geometrical shapes told the children that this was a 'school-type task', and so automatically flipped them into d-mode. They tried to figure out the rules, and couldn't. So they made no progress. The other version led them to reinterpret the display as a 'video game' – and this flipped them into an intuitive mode which enabled them to pick up the relevant relationships easily and unconsciously.[8]

Intuitions can also go wrong when they are based on inaccurate judgements about what is relevant and what is not, as we saw earlier with the 'mutilated chessboard'. Here is another example.

> A certain town is served by two hospitals. In the larger hospital about 45 babies are born each day, and in the smaller one, about 15. As you know, about 50% of all babies are boys and 50% girls. The exact percentage of girls however naturally varies from day to day. Some days it may be over the 50%; some days under. As a check on this variation, for a period of one year, both of the hospitals recorded the days on which more than 60% of the babies born were girls. Over the year, which hospital do you think recorded more such days? The large one? The small one? Or about the same?

When psychologists Daniel Kahneman and Amos Tversky asked nearly a hundred people this question, 22 per cent said the larger; 22 per cent said the smaller; and 56 per cent said 'about the same'.[9] Nobody sat down and worked it out with a calculator, so we must suppose that all these were intuitions. But more than three-quarters of them were wrong. (I was one of those who said 'about the same'.) A moment's reflection should be enough, however, to convince you – as it did me – that the correct answer is 'the small one'. The smaller the sample, the easier it is to get a larger percentage skew by chance. (It only takes two 'boys' to turn out to be girls for the small hospital to exceed its 60 per cent point.) A relevant piece of information – the size of the hospital – is actually being tacitly disregarded by half the population when they are generating their intuitive response (even though they are perfectly able to see its relevance when it is pointed out). These 'fast intuitions' are susceptible to all kinds of invisible influences, some of which will be appropriate and beneficial, and others of which will, in a particular instance, be misleading.

If fast intuition is vulnerable when responding to predicaments that look familiar but which are not as they seem, in what circumstances are the slower ways of knowing of most value? As with learning by osmosis, it turns out that slow intuition is good at uncovering non-obvious relationships between areas of knowledge; at seeing 'the pattern that connects' experiences that are superficially disparate. Intuition proves its worth in any situation that is shadowy, intricate or ill defined – regardless of whether the focus of concern is a mid-life crisis, a knotted-up relationship, an artistic project or a scientific conundrum.

In science, intuition is the faculty that comes up with the metaphor, the image or the idea that binds together and makes sense of experimental results which cumulatively seem to embarrass an existing theory, but which up to that point had lacked any alternative coherence. Both Darwin's account of the mechanism of evolution and Einstein's theories of special and general relativity offered just such explanatory patterns. They took a pile of details and transformed them into a theoretical structure that gave them meaning, and predicted new findings. And these, like many other scientific breakthroughs, came about through a way of knowing that was patient, playful and mysterious, not rational, earnest and explicit. As Einstein himself famously said, of his own creative process:

> The words of the language as they are written or spoken do not seem to play *any role* in my mechanism of thought. The psychical entities which seem to serve as elements of thought are certain signs and more or less clear images which . . . are in my case of visual and some of muscular type. [These elements take part in] a rather vague play . . . in which they can be voluntarily reproduced and combined . . . This combinatory play seems to be *the* essential feature in productive thought, before there is any connection with logical construction in words or other kinds of sign which can be communicated to others . . . In a stage where words intervene at all, they are, in my case, purely auditive, but they interfere [note, 'interfere'] only in a secondary stage. (Emphasis added)

Sometimes, as in the case of Herbert Spencer, one is aware of the pattern of thought gradually forming itself, as a large crystalline structure may slowly appear out of a saturated chemical solution in which a seed crystal has been placed. While at other times, the work proceeds unconsciously until the point at which the binding idea as a whole is delivered into consciousness. Rita Levi-Montalcini, who shared the Nobel Prize for medicine in 1986, for example, said: 'You've been thinking about something without willing to for a long time . . . Then, all of a sudden, the problem is opened to you in a flash, and you suddenly see the answer.' While Sir Neville Mott, physics laureate in 1977, confirms both the suddenness of the insight, and the difficulty of finding the right way of expressing it in d-mode: 'You suddenly see: "It must be like this". That's intuition . . . if you can't convince anybody else. This certainly happened to me in the work for which I got the Nobel prize. It took me years to get my stuff across.'[10]

Intuition may deliver its produce to consciousness in the form of more or less connected and coherent thoughts. But at other times, even for scientists, the undermind speaks in a variety of different voices. For Einstein, as for many creators, the language of intuition drew on visual and other forms of imagery. Kekulé first discovered that the carbon atoms of the benzene molecule linked up into a ring through watching the flames of his fire transform themselves, in his mind's eye, into snakes that turned round and bit their own tails. Sometimes intuition emanates in an almost aesthetic judgement: what Nobel chemistry laureate Paul Berg calls 'taste'. 'There is another aspect I would add to [intuition], and that is, I think, taste. Taste is almost the artistic sense. Certain individuals . . . in some undefinable way, can put together something which has a certain style, or a certain class, to it. A certain rightness to it.'

For others intuition manifests itself as a vague but trustworthy feeling of direction or evaluation – one 'just knows' which of several lines of enquiry to pursue, or which of a range of experimental results to take seriously, and which to ignore. Michael Brown (Nobel medicine laureate, 1985) describes how 'as we did our work, we felt at times that there was almost a hand guiding us. Because we would go from one step to the next, and somehow we would know which was the right way to go. And I really can't tell how we knew that . . .' While Stanley Cohen (Nobel medicine laureate, 1986), in similar vein, commented on the importance of developing a 'nose' for the important result – and of seeing this intuitive response as a valuable guide. 'To me it is a feeling of . . . "Well, I really don't believe this result", or "This is a trivial result" and "This is an important result" and "Let us follow this path". I am not always right, but I do have feelings about what is an important observation and what is probably trivial.' Note that Cohen acknowledges both the value and the fallibility of intuition. It can be wrong, and needs checking; but it none the less acts as source of guidance that is to be heeded and respected.

There are many accounts by creative artists and scientists of the need for patience and receptivity. In science, Konrad Lorenz, who won the Nobel Prize for medicine in 1973, stressed the importance of waiting. 'This apparatus . . . which intuits . . . plays in a very mysterious manner, because it sort of keeps all known facts afloat, waiting for them to fall in place, like a jigsaw puzzle. And if you press . . . if you try to permutate your knowledge, nothing comes of it. You must give a sort of mysterious pressure, and then rest, and suddenly BING, the solution comes.' While mathematician and philosopher George Spencer Brown declares, in his book *Laws of Form*:

> To arrive at the simplest truth, as Newton knew and practised, requires years of contemplation. Not activity. Not reasoning. Not calculating. Not busy behaviour of any kind. Not reading. Not talking. Not making an effort. Not thinking. Simply *bearing in mind* what it is that one needs to know.[11]

It is not, according to Lorenz and Spencer Brown, that one gives up on an intractable problem and drops it completely. The process is subtler than that. You do not try to figure it out, yet you 'give a sort of mysterious pressure'. You do not actively think, but you somehow 'bear the problem in mind'. It is as if you allow the problem to be there, to continue to exist on the edge of consciousness, yet without any purposeful attempt to bring it to a resolution. Nel Noddings, the American philosopher, describes this delicate balance of seeking and receiving in the more mundane context of studying a book.

> The mind remains, or may remain, remarkably active, *but instrumental striving is suspended*. In such modes we do not try to impose order on the situation but rather we let order-that-is-there present itself to us. This is not to say, certainly, that purposes and goals play no role in our submitting ourselves to a receptive state. Clearly they do. We may sit down with our mathematics or literature because we want to achieve something – a grade, a degree, a job – but if we are fortunate and willing, *the goal drops away*, and we are captured by the object itself.[12] (Emphasis added)

The gradual formation and development of an idea over a long time, perhaps from the tiniest of beginnings, and its delivery unwilled into consciousness, is a process that is as well known to artists as it is to scientists and mathematicians. Playwright Jean Cocteau both enthusiastically endorses the need to let the mind lie fallow, and attempts to scotch the idea that 'the muse' which springs from a patient state has anything magical or supernatural about it.

> Often the public forms an idea of inspiration that is quite false, almost a religious notion. Alas! I do not believe that inspiration falls from heaven. I think it rather the result of a profound indolence and of our incapacity to put to work certain forces in us. These unknown forces work deep within us, with the aid of the elements of daily life, its scenes and its passions, and, when . . . the work that makes itself in us, and in spite of us, demands to be born, we can believe that this work comes to us from beyond and is offered us by the gods. The artist is more slumberous in order that he shall not work . . . The poet is at the disposal of his night. His role is humble, he must clean house and await its due visitation.

The historian John Livingston Lowe has made a detailed study of the sources and materials on which Coleridge based 'The Ancient Mariner', and has been able to trace in these sources the forgotten antecedents of every word and phrase that appears in the most vivid stanzas.[13] He summarises the processes that must have been occurring, out of sight, in the poet's mind thus:

> Facts which sank at intervals out of conscious recollection drew together beneath the surface through almost chemical affinities of common elements . . . And there in Coleridge's unconscious mind, while his consciousness was busy with the toothache, or Hartley's infant ills, or pleasant strollings with the Wordsworths between Nether Stowey and Alfoxden, or what is dreamt in this or that philosophy – there in the dark moved the phantasms of the fishes and animalculae and serpentine forms of his vicarious voyagings, thrusting out tentacles of association, and interweaving beyond disengagement.[14]

Coleridge himself has described the composition of his other famous epic, 'Kubla Khan'. Feeling slightly 'indisposed', as he puts it, he took some opium, and settled down to continue his reading of a work called 'Purchas's Pilgrimage'. Shortly he dozed off, just as he was reading, 'Here the Khan Kubla commanded a palace to be built, and a stately garden thereunto. And thus ten miles of fertile ground were enclosed with a wall.' Three hours later he awoke 'with the most vivid confidence that he could not have composed less than two to three hundred lines – if that indeed can be called composition in which all the images rose up before him . . . without any sensation or consciousness of effort.' Immediately Coleridge grabbed pen, ink and paper and 'eagerly wrote down the lines that are here preserved'.[15]

American poet Amy Lowell describes how she uses incubation quite consciously as a trustworthy technique. 'An idea will come into my head for no apparent reason; "The Bronze Horses", for instance. I registered the horses as a good subject for a poem; and, having so registered them, I consciously thought no more about the matter. But what I had really done was to drop my subject into the subconscious, much as one drops a letter into the mailbox. Six months later, the words of the poem began to come into my head, the poem – to use my private vocabulary – was "there".'

Incubation is a process that may last for months or years, but its value is not confined to such long periods of gestation. It works over days (as when we 'sleep on it', and find the problem clarified, or even resolved, in the morning), or such short spans as a few minutes. The French mathematician Henri Poincaré, well known for his reflections on his own creative process, concluded:

> Often when one works at a hard question, nothing good is accomplished at the first attack. Then one takes a rest, longer or shorter, and sits down anew to the work. During the first half-hour, as before, nothing is found, and then all of a sudden the decisive idea presents itself to the mind . . . The role of this unconscious work in mathematical invention appears to me incontestable, and traces of it would be found in other cases where it is less evident . . .

There is now experimental evidence that corroborates these vivid anecdotes, and which helps us to understand how it is that incubation does its work. Steven Smith and colleagues at Texas A&M University have carried out a series of studies in which they were able to demonstrate incubation in the laboratory. Of course, they have not been able to reproduce the full complexity of the real-life creative experiences of an Einstein or a Coleridge. It is of the essence of such experiences that they cannot be directly manipulated or controlled. The undermind will not perform to order. Nevertheless, the results are informative.

The kinds of problems which Smith set his subjects were designed to mimic one of the key features of real-life creative insight: the discovery of a meaningful, but non-obvious, connection between different elements of the situation. So-called 'rebus' problems arrange words and images in such a way that they suggest an everyday phrase. For example:

ME JUST YOU

represents spatially the phrase 'just between me and you'. Or

TIMING TIM ING

is a visual pun on the expression 'split second timing'.

Subjects were shown a succession of such puzzles, and initially given thirty seconds in which to attempt to solve each one. Some of the puzzles were accompanied by helpful 'clues' (such as 'precise' for the second example above), or unhelpful ones (such as 'beside' for the first one). Those problems that the subjects failed to solve the first time round were re-presented for a second try either immediately, or after a delay of five or fifteen minutes. When they had a second go immediately, subjects showed no improvement over their initial score. But when they were retested after a delay, performance improved by 30 per cent on the puzzles that had been accompanied by the unhelpful clues; and the longer (fifteen-minute) delay produced greater improvement than the shorter (five-minute) one. Significantly, the improvement did not depend on whether subjects had been able to work consciously on the problems during the delay, or had been given an irrelevant task to occupy their attention. So the benefit of incubation in this situation cannot be explained on the basis of having longer to think purposefully.

In another study, Smith elicited the incubation effect using the 'tip-of-the-tongue' (TOT) phenomenon, which occurs when you are trying to recall something – a name, typically – which just won't come to mind, though you have the strong feeling that it is 'on the tip of your tongue'. Using computer graphics, Smith invented pictures of imaginary animals, to which he attached names and a brief description of their supposed habits, habitats and diets. Subjects were given twelve such animals to study briefly, and then were asked to recall their names. As in the previous study, for the names they were unable to remember they were given a second recall test either immediately or after a five-minute delay. On this second test, they were asked to indicate, if they still could not get the name, what the first letter might be, if they thought they would recognise the name if they were shown it, and whether they felt that the name was on the tip of their tongue or not. The delay improved memory by between 17 per cent and 44 per cent And furthermore, even if subjects were unable to recall the whole name, their 'guesses' as to the first letter were more accurate when they said they were in the TOT state.

Smith suggests that in both studies there is a common explanation for the positive effect of incubation: the delay allows time for wrong guesses and blind alleys to be forgotten, so that when you come back to the task, you do so with a more open mind. There is a tendency to get fixated on a particular approach, even when it is patently not working. The delay increases the chances that your mind will stop barking up the wrong tree. 'When the thinker makes a false start, he slides insensibly into a groove and may not be able to escape at the moment. The incubation period allows time for an erroneous set to die out and leave the thinker free to take a fresh look at the problem.'[16]

The idea that delay encourages a release from fixation, that it enables you to shake off unpromising approaches or assumptions that have been blocking progress, is certainly one aspect of incubation, but it cannot be the whole story, for it fails to take into account the active workings of unconscious intelligence. The fact that we can tell with a fair degree of accuracy when we are in the TOT state, whether we would be able to recognise the name if it were shown to us, and even, sometimes, retrieve its initial letter, or some other characteristic such as the number of syllables, suggests that the undermind has an idea what the word is, but for some reason is not yet willing or able to release that hypothesis fully into consciousness.

Yaniv and Meyer have shown directly that this sort of subliminal knowledge does exist. Like Smith, they investigated the TOT effect, this time reading to their subjects definitions of rare words and collecting those that subjects could not recall but felt sure they knew. They then used these words, along with other new words, in a 'lexical decision task', in which strings of letters were flashed on to a computer screen, and subjects had to press one of two keys to indicate, as quickly as they could, whether the string was a real word or not. It has been shown previously that words which have recently been seen, prior to the test, are recognised as being real words faster than other words which are equally familiar but which have not been recently 'activated' in memory. Yaniv and Meyer found that, even though the TOT words had not been *consciously* recalled, they still showed this 'priming' effect, indicating that they had been activated in memory, despite the fact that the 'strength' of the activation had not been great enough to exceed the threshold required for consciousness.[17]

One of the effects of this partial activation is to increase the likelihood that some chance event may provide the extra little 'nudge' that is needed to get the word to tip over the threshold into consciousness – and this provides another way in which incubation can come about. Consciously you may think that you have made no progress towards the solution of the problem, and may even feel that you have given up. But unconsciously some progress might have been made; not enough to satisfy the criteria for consciousness, but enough to leave the 'candidate' somewhat primed. If some random daily occurrence serves to remind you, even if only subliminally, of the same word or concept, that may be sufficient to tip the scales, and you have the kind of sudden, out-of-the-blue experience of insight to which personal accounts of creativity often refer. Many people have had the experience of suddenly remembering a dream during the course of the day, when some trivial stimulus, such as a fragment of overheard conversation, acts as a sufficient trigger for conscious recollection.

In the discussion of 'learning by osmosis', we saw that the undermind may be making progress in picking out a useful pattern of which the conscious mind is unaware. In such situations, we can show that we know more than we think we know. Does the same apply to the kinds of problem-solving that we are looking at in this chapter? Can we demonstrate directly that the undermind is closer to the solution of a problem than we think? And can we learn to be more sensitive to the subtle clues or indications that this is the case? Should we place greater trust, perhaps, in ideas that just pop into our minds, rather than treating them as random noise in the system, to be ignored? Recent studies by Kenneth Bowers and his colleagues at the University of Waterloo in Canada have suggested positive answers to these questions.

Like Smith, Bowers assumes that intuition is closely related to the ability to detect the underlying link or pattern that makes sense of seemingly disparate elements, and he has used both visual and verbal stimuli to explore the ways in which the undermind can home in on such patterns before conscious, deliberate thought has any idea what is going on. Consider the images shown in Figure 5.

One of each pair of pictures, either A or B, shows a highly degraded image of a real object.[18] The other shows the same visual elements arranged in a different configuration. Subjects were shown a series of such pairs, and asked to write down the name of the object depicted in one of the drawings. If they were unable to do so, they were asked to guess which of the two images actually represented the real object, and to indicate their degree of confidence in this 'guess'. The results showed that people's guesses were better

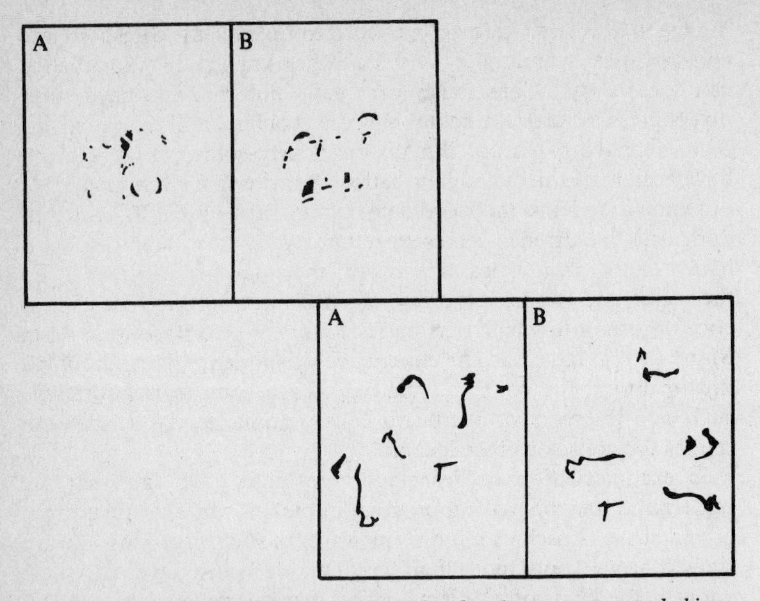

Figure 5. Bowers' degraded images. One of each pair represents a real object: a camera (top) and a camel (bottom)

than chance, and that this was so even when they indicated that they had zero confidence in their guess. The possibility that the visual fragments of the real objects were somehow more coherently arranged, and that it was this perceptual clue, rather than any unconscious activity, that was informing the guesses, was discounted by showing the pairs of shapes to naïve subjects and asking them to rate directly which looked the more 'coherent'. There was no difference in this judgement between the real and the rearranged images. Thus it appears, just as with the tip-of-the-tongue state, that the unconscious is able to indicate, in the form of what consciousness judges to be a complete guess, that it has got some way towards detecting the pattern, even when it has not yet been unequivocally identified.

The same finding is obtained with verbal rather than visual stimuli. Below are three pairs of three words. Within each pair, one of the sets of three words has a common (but not very obvious) associate – a single word that is related in some way to each of the three – while the other set of words is not connected in this way.[19]

	A	B
1	STICK	PARTY
	LIGHT	ROUND
	BIRTHDAY	MARK
2	HOUSE	MAGIC
	LION	PLUSH
	BUTTER	FLOOR
3	WATER	SIXTEEN
	TOBACCO	SPIN
	LINE	TENDER

As with the pictures, people were asked to try to spot the connection, and if they could not, to indicate which of the two sets was the one that *did* have the (undetected) link. The results were essentially the same as with the visual stimuli: people were able, some of the time, to detect the presence of a pattern that they could not identify, and were able to do so more reliably than their own confidence ratings would suggest.

In an ingenious elaboration of this last study, Bowers devised what he called the 'accumulated clues task'. The problem was similar to the one just described: subjects had to discover the single word that formed the common associate to a number of other words. But now there was a list of fifteen such words, and they were presented in sequence, one at a time, rather than all together.[20]

Accumulated Clues Test

1. RED
2. NUT
3. BOWL
4. FRESH
5. PUNCH
6. CUP
7. BASKET
8. JELLY
9. COCKTAIL
10. GUM
11. PIE
12. TREE
13. BAKED
14. SALAD
15. FLY

The first word was presented for ten seconds or so, during which time subjects were obliged to write down at least one association. Then the second cue word was revealed, and another response was required; and so on. When subjects thought that they had found a response that was a viable hunch or hypothesis, they marked it, but continued making further responses until they either changed their mind or were convinced that they had found the target word. Typically, over a number of such tests, people found a viable candidate on about the tenth word, and settled on a firm answer after receiving about twelve of the cue words.

If the unconscious can run ahead of consciousness, then people's 'guesses' might begin to converge on the target word before they realise it. In order to check this, the responses that subjects gave *before* they settled on a plausible hypothesis were presented to a panel of judges, to see if they bore any meaningful relationship to the as yet unidentified target word. Sure enough, they did. If they looked back over people's guesses, independent judges, who knew the solution, could see a pattern of steady approximation to the target; a pattern of which the subjects themselves were unaware. It appears that the ideas that just pop into our heads may have greater validity than we think, and that we therefore deprive ourselves of useful information if we ignore them, or treat them as 'complete guesses'.

Bowers himself notes one important way in which these stylised problems are unrepresentative of problem-solving in real life. In the real world, a major part of the 'problem' is often that one does not know in advance what is relevant and what is not. The predicaments confronted by a business executive, an architect, a research scientist or a teacher are 'messy', in the sense that it is often not at all clear, at the beginning, how to conceptualise the problem, or what aspects of the available information to pay attention to and what to discard. The novice driver or medical student frequently feel overwhelmed with data because they have not yet discovered through experience what matters and what does not; what needs placing in the foreground of awareness and what can recede into the background. Bowers' puzzles, like many of those used by psychologists (and by those who design school curricula), are carefully tidied up before they are presented. The image of the camera is degraded, but there is no 'noise' in it. So in his most recent experiments, Bowers has made his problems more messy, and more lifelike. His tests now include some information that is irrelevant or distracting, as well as information that is relevant and valuable. Similar results seem to be emerging. For example, as they get closer (unbeknownst to themselves) to a solution, so subjects get better at 'guessing' which of the details of the problem are actually relevant.

We now have empirical evidence for the existence of the mysterious 'guiding hand' that told Nobel laureate Michael Brown which step to take next, and Stanley Cohen which result to 'believe' and which to doubt. There is evidence, in other words, for the undermind, the intelligent unconscious that works quietly below, and in some cases ahead of, conscious apprehension. Both poets and scientists have always suspected as much. If they are observant, as they must be, they know it through their direct experience. When Amy Lowell was asked, 'How are poems made?', she replied: 'I don't know. It makes not the slightest difference that the question as asked refers solely to my own poems, for I know as little of how they are made as I do of anyone else's. What I do know about them is only a millionth part of what there must be to know. *I meet them where they touch consciousness, and that is already a considerable distance along the road of evolution.*' (Emphasis added)

While R. W. Gerard, writing in *The Scientific Monthly* in 1946, foreshadowed, with his acute speculations, much of what cognitive science is finally beginning to demonstrate beyond question.

> Much attention has been given to the phenomena of learning: by the slow cumulation of a correct response in the course of experience ['learning by osmosis']; and by the sudden grasp of a solution and abrupt performance of the correct response ['intuition']. They seem very different . . . but it is possible, perhaps probable, that they are basically quite similar. In both cases, new functional connections must be established in the brain; and this process may be more gradual and cumulative in the case of 'insight' than it appears. For here, also, much brain work precedes the imaginative flash – the theory of gravitation may result only when the metaphorical apple falls on the prepared mind – and only when the process has progressed to some threshold level does it overflow into a conscious insight.

CHAPTER 5

Having an Idea: the Gentle Art of Mental Gestation

You cannot go into the womb to form the child; it is there and makes itself and comes forth whole . . . Of course you have a little more control over your writing than that; but let it take you and if it seems to take you off the track don't hold back.

Gertrude Stein

There are a number of metaphors that creators use to describe their process, but none more common than that of gestation. 'Having' a good idea is akin, they say, to having a baby. It is something that needs a seed to get started. It needs a 'womb' to grow in that is safe and nurturing, and which is inaccessible. The progenitor is a host, providing the conditions for growth, but is not the manufacturer. You 'have' a baby, you do not 'make' it – and so with insight and inspiration. Gestation has its own timetable: psychologically, as biologically, it is the process *par excellence* that cannot be hurried. And it cannot be controlled; once the process has been set in motion it happens by itself, and will, barring any major accident or intervention, carry through to fruition.

It is not just romantics who see the mind this way. Even the arch-behaviourist B. F. Skinner once gave a lecture at the Poetry Center in New York which he entitled 'On "Having" a Poem', and which he started by explaining that his talk had the curious property of illustrating itself, in that he was at that moment in the throes of 'having' a lecture.[1] And he went on to develop the metaphor in more detail. 'When we say that a woman "bears" a child, we suggest little by way of creative achievement. The verb refers to carrying the foetus to term.' And then, after she has 'given' birth to the child

– as if birth were some kind of property or gift that can be bestowed – we tend to say merely that she has 'had' a baby, where 'had' can seem to mean little more than 'came into possession of'.

What precisely is the nature of the mother's contribution? She does not decide upon the colour of the baby's eyes or skin. She gives it her genes, but are they really 'hers', when she inherited them from *her* parents, and through them from an entirely unwilled lineage? She surely cannot take much personal pride in the hazel eyes and the auburn hair she is handing on. 'A biologist', says Skinner, 'has no difficulty in describing the role of the mother. She is a place, a locus in which a very important biological process takes place. She provides warmth, protection and nourishment, but she does not design the baby who profits from them. The poet is also a locus, a place in which certain genetic and environmental causes come together.' And, as we have seen, what is true of the poet can be equally true of the scientist, the novelist, the sculptor or the product designer.

The analogy reminds us that, although the process of creativity is essentially organic rather than mechanical, nevertheless the nature of the 'incubator' is vital to the germination of the seed. The mother does not engineer her child's intrauterine development, but she influences it enormously through her lifestyle and her sensitivity, her anxieties, appetites and attitudes, her history and her constitution. Who she is, and the physical and emotional environment that she herself inhabits, affects the nature and the quality of the sanctum that she provides for the growing form of life within her. And so it seems to be with intuition: there are conditions which render the mental womb more or less hospitable to the growth and birth of ideas; and differing ways in which, and extents to which, different people are able, wittingly or unwittingly, to provide those conducive conditions. The more clearly we can identify what these conditions are, the more able we shall be to see how they can be fostered.

First, one needs to find the seed – and this process, for the creator, requires curiosity: an openness to what is new or puzzling. One must allow oneself to be impregnated. Unless one is piqued by a detail that obstinately refuses to fit the conventional pattern, or a chance remark that somehow resonates with one's own unexplicated views or feelings, there is nothing for the creative process to work upon. A. E. Housman breaths life into a hackneyed image when he says: 'If I were obliged to name the class of things to which [poetry] belongs, I should call it a secretion; whether a natural secretion, like turpentine in the fir, or a morbid secretion, like the pearl in the oyster. I think that my own case, though I may not deal with the material so cleverly as the oyster does, is the latter.'[2]

For scientists, the stimulus is often an unexplained detail or incongruity. The imaginative seed that finally flowered in the theory of relativity was the teenage Einstein's attempt to imagine what it would be like to ride on a beam of light. While making a routine check through three miles of computer print-out from the radio telescope, a young Cambridge astrophysicist spotted just a few traces that puzzled her. They could easily have been ignored, or written off as noise. But, with a lot of subsequent hard work, this observation finally resulted in the discovery of a completely new type of star. Out of hundreds of tiny fruit flies, one had a misshapen eye. A biologist could not help wondering why. Five years later, his investigations have led him to the discovery of a kind of receptor protein that may well be implicated in the production of cancer cells.[3]

In the commercial world, the competitive edge belongs to the executive or product developer who is capable of sensing the potential in an apparent setback, or taking time to mull over the meaning of a quirk in the market. And the reflective accounts of artists, too, reveal the importance of this sensitivity to poignant trifles. In the preface to his story 'The Spoils of Poynton', Henry James explains how essential such details are. One Christmas Eve he was dining with friends when the lady beside him made, as he puts it, 'one of those allusions that I have always found myself recognising on the spot as "germs" . . . Most of the stories straining to shape under my hand have sprung from [such] a . . . precious particle. Such is the interesting truth about the stray suggestion, the wandering word, the vague echo, at the touch of which the novelist's imagination winces as at the prick of some sharp point: its virtue is all in its needle-like quality, the power to penetrate as finely as possible.'[4]

It seems that such seeds implant themselves only in those who at an unconscious level are already prepared. Even if the issue is an intellectual rather than an artistic one, its recognition is personal, affective, and even aesthetic (such as Nobel laureate Paul Berg talking about how important, in his work, was the sense of 'taste' for a problem or an approach). Novelist Dorothy Canfield, in the same vein as Henry James, recounts the incident that formed the nucleus of her story 'Flint and Fire'. She had some business with a neighbour, and to get to his house had to walk along a narrow path through dark pines, beside a brook swollen with melted snow. Emerging from the wood, she found the old man sitting silent and alone in front of his cottage. Having done her business, and keen not to offend against country protocol, she sat beside him to chat for a few minutes.

> We talked very little, odds and ends of neighbourhood gossip, until the old man, shifting his position, drew a long breath and said, 'Seems to me I never heard the brook sound so loud as it has this spring.' There came instantly to my mind the recollection that his grandfather had drowned himself in that brook, and I sat silent, shaken by that thought and by the sound of his voice . . . I felt my own heart contract dreadfully with helpless sympathy . . . and, I hope this is not as ugly as it sounds, I knew at the same instant that I would try to get that pang of emotion into a story and make other people feel it.[5]

Stephen Spender said that his experience of inspiration was that of a 'line or a phrase or a word or sometimes something still vague, a dim cloud of an idea which I feel must be condensed into a shower of words'.

So the seed will not germinate unless it makes contact with a 'body of knowledge' of the right kind, in a congenial state. But what exactly is the 'right' kind? The evidence from studies of conspicuous innovators suggests that this pre-existing body is most fecund when it is full of rich experience – but not to the point where it has become so familiar that it is automated and fixed. One must have the evidence on which to draw, and one must know enough to be able to recognise a good idea when it comes. Clearly one cannot be creative *in vacuo*. But if one is too steeped in the problem, the danger is that the grooves of thought become so worn that they do not allow a fresh perception, or a mingling of different currents of ideas, to occur. Recall the experiments with the water jars, and the fact that people quite quickly became 'set in their ways'. The more experience they had had with the complex rule, the less likely they were to spot the simpler solution when it became available. Studies of creative individuals generally show an inverted U-shaped relationship between creativity and age. In mathematics and the physical sciences, for example, the age of peak creativity is between twenty-five and thirty-five.[6]

To give a more specific example: the *New York Times* carried a front-page article on 18 February 1993 reporting the discovery of the first successful technique for eliminating the AIDS virus from human cells *in vitro*, and also for preventing the infection of healthy cells. The inventor of this method was a medical graduate student,

Yung Kang Chow, who, precisely because of his relative inexperience, was able to see through a blocking assumption which researchers had, up to that point, been unconsciously making. Chow speculated: 'Perhaps by virtue of being a graduate student and not having learned much medicine yet, I had more naive insight into the problem.' Seeing through an existing, invisible assumption, which is often the key to creativity, requires a mind that is informed but not deformed; channelled but not rutted.[7]

Intuition, as we have seen, tends to work best in situations that are complex or unclear, in which the information that is given may be sketchy or incomplete, and in which progress can only be made by those who can, in Jerome Bruner's famous phrase, 'go beyond the information given', and are able to draw upon their own knowledge in order to develop fruitful hunches and hypotheses. Both novelist and scientist may well need to go out and collect more 'data', but the creative idea comes from bringing into maximum contact the 'problem specification', the data, and one's own store of experience and expertise; allowing these to resonate together as intimately and as flexibly as possible, so that the full range of meaning and possibility of both current data and past experience are extracted. The good intuitive is the person who is ready, willing and able to make a lot out of a little.

If you insist on having high-quality information from impeccable sources before you are willing to form a judgement, you may reduce the occasions on which you are obviously 'wrong'. You will make instead 'errors of omission' that are often less visible. By adopting such a conservative attitude, you may also fail to make use of the more tentative, holistic responses that are authorised by the unconscious. On the other hand, if you are indiscriminately intuitive, you are more ready to back hunches on the faintest of whims. The crucial question concerning intuition, therefore, is how to relate to conscious and unconscious in such a way that *both* kinds of mistake are minimised; so that you are open to the promptings of the undermind, willing to hear and acknowledge them, yet not over-respectful or lacking in discernment.

Are people differentially willing to make judgements and decisions on the basis of inadequate (conscious) information? And if so, of those who are willing, are some better at it than others? Studies by Malcolm Westcott at Vassar College in America show that the answer to both these questions is a clear 'yes'. Westcott gave his undergraduate subjects one example of a relationship that could hold between two words or two numbers, and their task was to show that they had discovered the rule or relationship by adding the correct 'partner' to another word or number. So they might be shown '2, 6', and asked to complete the pair '10, ?', or 'mouse, rat', and then 'weekend, ?'. But subjects were also given other sealed clues, which they could ask to be revealed, one by one, before they were ready to give their answer. They were free to look at as many or as few of these other clues as they wished before responding. When the students gave their answers, they were also asked to rate how confident they were that they were in fact correct. Westcott was thus able to take three measures for each problem: whether the solution was right or wrong; how many clues people wanted to see before they gave their response; and how confident they were. He repeated the experiment with several different groups of subjects both in England and America, and with different kinds of problems.

He found that people differ markedly, and consistently, on all three measures of their performance, so much so that it was possible to identify four quite different sub-groups. There were those who typically required very little information before offering their solutions, and who were likely to be correct. These he called the 'successful intuitives'. Then there were those who also took little extra information, but who tended to be wrong – the 'wild guessers'. The third group required a lot of information before being willing to respond, but were generally successful when they did: the 'cautious successes'. And finally there were those who made use of all the information they could lay their hands on, but who still made a lot of mistakes, the 'cautious failures'.

Westcott also gave various personality tests to his subjects, so that he was able to see what the characteristics of the successful intuitives (and of the other groups) were. He found that the good intuiters tend to be rather 'introverted'; they like to keep out of the social limelight, but feel self-sufficient and trust their own judgement. They like to make up their own minds about things, and to resist being controlled by others. They tend to be unconventional, and comfortable in their unconventionality. In social gatherings they are 'composed', but are capable of feeling strongly, and showing their feelings in more intimate or solitary situations. They enjoy taking risks, and are willing to expose themselves to criticism and challenge. They can accept or reject criticism as necessary, and they are willing to change in ways they deem to be appropriate. They describe themselves as 'independent', 'foresighted', 'confident' and 'spontaneous'. '*They explore uncertainties and entertain doubts far more than the other groups do, and they live with these doubts and uncertainties without fear.*'[8]

In contrast, the 'wild guessers' are much more socially orientated. But 'their interactions are characterized by considerable strife, they seem to be quite self-absorbed, and their "affective investments" seem to be directed towards themselves'. These traits often manifest themselves as 'a driven and anxious unconventionality, coupled with strong and rigid opinions, and overlaid with cynicism . . .' They describe themselves as 'alert', 'quick', 'headstrong' and 'cynical'. Westcott comments that these people 'appear to be striving for a grasp of reality which so far eludes them, and they are likely to attempt different modes of attack [on uncertainty] in a somewhat chaotic manner'.

The 'cautious successes' are distinguished by 'a very strong preference for order, certainty and control', and they have a high respect for authority. They are well-socialized, in the sense that their stated interests and values are in the mainstream of their culture, but they do not recognize that they have been influenced in this. Their desire for certainty and order seems to lead them to some social awkwardness and anxiety in the uncertain world of interpersonal relations. Affect is difficult for them to handle, unless it is very well structured, and they describe themselves as 'cautious', 'kind', 'modest' and 'confident'. The overall picture of this group, according to Westcott, is 'one of conservative, cautious, somewhat repressive people who function well in situations where expectations are well-established and well met': d-mode types, we may assume.

Finally, the 'cautious failures' have a view of the world 'in which everything is risky at best, and they are essentially powerless to influence or control it. There is a broadly generalized passivity, a sensitivity to – and felt inability to deal with – injustice, and a wish for a quiet, certain *status quo*; through all of this they lack self-confidence . . . They are quite conservative, presumably as the best defence against the great uncertainties of life, and they seem to wander through life just managing to keep their heads above water, not making waves. They see themselves as "cautious", "kind" and "modest".'

Perhaps the most significant of all these interesting findings is that the group who are most at ease with uncertainty and doubt, the most able to 'live with it', are the group who are most able to make successful use of the inadequate information they have. They can use their unconscious resources to help them make *good guesses* in uncertain situations, and are willing to do so. This provides some powerful empirical corroboration for the idea that the flight from the experience of uncertainty pushes people into the exclusive use of a cognitive mode which is ill suited to dealing with some puzzling situations. It is also very significant that, having reviewed the relevant studies to date, Westcott concludes that 'intuition is most likely to occur when the information on which an inference is based is excessively complex, apparently absent or limited, or when the time necessary for explicit manipulation of data is not available . . . These are all conditions which remove the thinker from direct application of adult, socially validated logic.'[9] American social scientist Donald Schon has recently argued that it is situations of just this sort that are routinely faced by professionals such as teachers and lawyers.[10] Although there are bodies of helpful precedents and maxims, such people spend much of their time dealing with cases that are sufficiently unique, and sufficiently complex, to prevent the straightforward application of any rule-book. They are off the well-laid-out highways of 'technical rationality', trying to find their way through what Schon refers to as the 'swampy lowlands' of professional practice.

Sometimes this resonating of data and experience – perception and cognition – happens quickly. Westcott's puzzles are sufficiently simplified and stylised to allow intuition to work quite rapidly. Not much experience has to be brought to bear. No remote analogies or metaphors need to be found. No very subtle patterns connecting apparently disparate elements have to be uncovered. Very often, though, when the predicament is more intricate, the undermind needs to be left to its own devices for a while, and then the need for patience – the ability to tolerate uncertainty, to stay with the feeling of not-knowing for a while, to stand aside and let a mental process that can neither be observed nor directed take its course – becomes all important.

Someone who cannot abide uncertainty is therefore unable to provide the womb that creative intuition needs. Milton Rokeach, having, as we saw in Chapter 4, showed that creativity is enhanced when people are forced to slow down, concludes that 'differences between people characterised as rigid, and others characterised as less rigid, may be attributable . . . to personality differences in time availability . . . Time availability [i.e. the willingness to think slowly] makes possible broader cognitions, more abstract thinking . . . and consequently greater flexibility.' And he goes on to offer a plausible speculation as to how these differences may arise. 'Some individuals, because of past experiences with frustrating situations involving delay of need satisfaction, become generally incapable of tolerating frustrating situations. To allay anxiety, such individuals learn to react relatively quickly to new problems . . . The inevitable consequence is behavioural rigidity.'[11] Whether one is or is not a good intuiter therefore turns out to be a matter of cognitive habits or dispositions – but these are underpinned by emotional and personal characteristics that may be quite deep-seated. If one is threatened by the experience of ignorance, then one cannot dare to wait, and may, as a result, cling to a mode of cognition – d-mode – that is purposeful and busy, seeming to offer a sense of direction and control, which may be the wrong tool for the job in hand.

There is a wealth of evidence to confirm the common impression that when people feel threatened, pressurised, judged or stressed, they tend to revert to ways of thinking that are more clear-cut, more tried and tested and more conventional: in a word, less creative. Studies with the Luchins' water jars problem have shown that adherence to the over-complicated solution, when an easier one becomes available, is increased by stress. In an old study (which would certainly not be approved these days by an ethics committee), students were told that, on the basis of a previously administered questionnaire, there was evidence that they possessed some 'maladapted personality features', and that their performance on the water jars problem would clarify the situation. The more threatened the subjects felt, the more tenaciously they clung to the outdated solution, and the less likely they were to spot the new possibility.[12]

Less severe degrees of stress also disrupt performance. Arthur Combs and Charles Taylor gave people the task of encrypting some sentences according to the kind of simple transposition code that one finds in spy stories in children's comics. Some of the sentences were 'incidentally' of a personal nature, such as 'My family do not respect my judgement', while others were neutral ('The campus grew quite drab in winter'). Some of the neutral sentences were preceded by the experimenter saying mildly, 'Can't you do it a little bit faster?' The 'personal' sentences tended to be encoded more slowly, and with more mistakes, but the worst condition of all was when a neutral sentence was accompanied by the time pressure. Even in such a straightforward task, where the degree of creativity required is minimal, the exhortation to 'hurry up' is entirely counterproductive.[13]

The deleterious effect of time pressure on the quality of thinking is also shown in a study by Kruglansky and Freund. Students were given some personal data about a hypothetical applicant for a managerial job and asked to predict his likely success in the position. Half the students were given positive information followed by negative, and the other group were given the same information in the reverse order. Those students who received the positive information first gave significantly higher predictions of success than the others. And this tendency was exaggerated when the students were asked to make their judgements against the clock. What seems to happen is that we build up an intuitive picture of the situation as we go along, and it takes work to 'dismantle' this picture and start again. So if later information seems to be at odds with the picture so far, we may unconsciously decide to reinterpret the dissonant information, rather than radically reorganise the picture. And the more we feel under pressure, the less likely we are to make the investment of 'starting from scratch'. This tenacity is a considerable pitfall for intuition, for when we are making real-life decisions it often happens that information is not available to us all at once, but arrives piece by piece. If we 'make up our minds' quickly and intuitively, it means that later pieces of information may be ignored or downgraded if they do not happen to confirm the judgement that has already been made.[14]

Even stress that is not particularly related to the problem-solving task itself increases this rigidity. Hospital patients who are awaiting an operation give more stereotyped responses than control subjects to the Rorschach ink-blot test, and are much less fluent and creative in thinking up ways to complete similes such as 'as angry (or "interesting" or "painful") as —'. They also, incidentally, become physically more clumsy and more forgetful.[15]

One of the people who has worked most intensively on increasing the quality of intuition in practical, real-life settings is George Prince, the founder, with William Gordon, of the well-known 'Synectics' programme for enhancing creativity. Prince started out with the assumption that people needed to be trained in the art of generating more and better ideas. 'I was convinced that people tended to come to us weakly creative and leave strongly creative.' But slowly he became convinced that this was not the nub of the problem. He realised that *speculation*, the process of expressing and exploring tentative ideas in public, made people, especially in the work setting, intensely vulnerable, and that, all too frequently, in a variety of subtle (or not so subtle) ways, people came to experience their workplace meetings as unsafe.

People's willingness to engage in delicate explorations on the edge of their thinking could be easily suppressed by an atmosphere of even minimal competition and judgement. 'Seemingly acceptable actions such as close questioning of the offerer of an idea, good-natured kidding about someone's idea, or ignoring the idea – any action that results in the offerer of the idea feeling defensive – tend to reduce not only his speculation but that of others in the group.' Prince's depressing conclusion is that adults in the workplace are much more susceptible to 'hurt feelings' than we commonly admit, and that equally prevalent is the largely unacknowledged tendency for workers at all levels, and in all vocations, to see themselves as engaged in a competitive struggle to preserve and enhance a rather fragile sense of self-esteem. He concludes: 'The victim of the win–lose or competitive posture is always speculation, and therefore idea production and problem solving. When one speculates he becomes vulnerable. It is too easy to make him look like a loser.'[16]

Just as mothers-to-be may become rather particular about the conditions in which their gestation, and eventually the birth, takes place – traditionally demanding special foods or comforts that may seem to others somewhat eccentric – so too do creators, according to their own testimony, sometimes develop personal rituals and requirements that establish the conditions which are felt to be safe and conducive to intuition. Pearl S. Buck could not work without a vase of fresh flowers on her desk, and a view of the New England countryside, while Jean-Paul Sartre hated the country, and needed to look out on to the bricks and chimneys of a Parisian street. Kipling claimed to be unable to write anything worthwhile with a lead pencil. The poet Schiller liked to fill his writing desk with rotting apples, claiming that the aroma stimulated his creativity. Walter de la Mare, Sigmund Freud and Stephen Spender, along with many others, had to chain-smoke while writing. Though collective 'brainstorming' is valuable for throwing up novel ideas, the conditions for deeper insight and intuition seem most often to be solitary and free from outside pressure of any kind. Carlyle tried to build a soundproof room. Emerson would leave home and family for periods and live in a hotel room. I can do three months' good work in a fortnight in my New Zealand beach-house.

It is not only a hostile external environment that can reduce creativity. If our own belief systems are threatened, by, for example, some unanticipated implications that seem to be emerging from a seemingly innocuous line of thought, we may suppress our own intuition and speculation. What started out as an intriguing puzzle may, as we dig deeper into it, turn out to have unwanted repercussions for the way we think about and organise our lives. The more fundamental a belief is to our view of ourselves, or to a position on which we have staked our reputation, the harder it is going to be to re-examine. Sometimes this mental inertia is entirely reasonable: wholesale reorganisation of the mental household is not to be undertaken lightly. If someone suggests that we rearrange the furniture, so to speak, we might be willing to try it; but if they propose that the house would look better if we were to move the foundations a few metres to the right, they are likely to meet rather stronger resistance. Just so with fundamental changes to the structure of our knowledge.

Efraim Fischbein from Tel Aviv University comments on the inertia of science in this regard, but the same principle applies to the informal, everyday mind just as well.

> A scientist who has formulated a certain hypothesis did not formulate it by chance; it optimally suits his general philosophy in the given domain, his usual way of interpretation, his knowledge and his research methodology. He is certainly very anxious to preserve his initial interpretation not only for his own prestige – which is certainly an important factor – but chiefly because it is the hypothesis which is best integrated in the structure of his reasoning. He will be unwilling to give up this first hypothesis because by renouncing it he has to re-evaluate a whole system of conceptions.[17]

Hence what has come to be referred to as 'Planck's dictum', after the German physicist Max Planck: major advances in science occur not because the proponents of the established view are forced by the weight of evidence to change their minds, but because they retire and eventually die.

It is not only the whole class of things that we refer to as 'threats' which militates against the relaxed and hospitable mood that encourages creativity: anything that simply makes you try too hard has the same effect. Wanting an answer too much can interfere with the process of gestation. In one study, Carl Viesti asked his subjects to try to detect which of three complicated patterns was the odd one out, and looked at the extent to which their performance improved over a series of such tests. Although they were given plenty of time to examine each set, those subjects who were offered significant rewards for correct detection performed worse, and learnt less, than those who were given only a token payment. Viesti concludes that 'regardless of their size, monetary utilities [sic] do not appreciably increase performance on insight learning tasks, rather, their presence may interfere with such performance'.[18]

Interestingly, the same counterproductive effect of incentives has also been observed in the animal world. Rats and monkeys who have to learn a skill in order to get food discover less about their environment in general if they are ravenous than if they are only mildly hungry. The more pressing is the requirement to reach the goal or solve the problem, the less do animals or human beings attend to the overall patterns in their world, and the more they try to pick out just those few pointers that will get the job done. This is adaptive up to a point; but if the world then changes, so that there are new contingencies to be discovered, such an attitude is exposed as blinkered and narrow.[19] Incentives may increase routine productivity, it seems; but they do not create conditions conducive to top-quality insights and solutions. Too many carrots, as well as too much stick, are inimical to creative intuition.

The next quality which encourages creative intuition we might call 'feeling it kick'. As the seed of an idea grows, it is as if the host gradually becomes aware of the autonomous movements of new creative life inside her. And how sensitive she is to these small signals, and how she responds to them, has a significant influence on the creative process. How mental gestation turns out depends particularly on the ability to turn on to the borderlands between consciousness and the unconscious a kind of awareness that is welcoming without being predatory, and perceptive without being blinding. Crucially, skilled intuiters seem to be able to watch the emergence of their creations without chivvying them, neatening them up or trying to turn them too quickly into words.

In the 1960s poet Ted Hughes gave a series of talks for young people on the radio about writing. In one of these, he described very beautifully this quality of gentle attentiveness to one's own mind.

> At school . . . I became very interested in those thoughts of mine that I could never catch. Sometimes they were hardly what you could call a thought – they were a dim sort of a feeling about something . . . [and] for the most part they were useless to me because I could never get hold of them. Most people have the same trouble. What thoughts they have are fleeting thoughts – just a flash of it, then gone – or, though they know they know something, or have ideas about something, they just cannot dig those ideas up when they are wanted.

> Their minds, in fact, seem out of their reach . . . The thinking process by which we break into that inner life . . . is the kind of thinking we have to learn, and if we do not somehow learn it, our minds lie in us like fish in the pond of a man who cannot fish . . . Perhaps I ought not to call it thinking at all. I am talking about whatever kind of trick or skill it is that enables us to catch these elusive or shadowy thoughts, and collect them together, and hold them still so we can get a really good look at them.[20]

Hughes goes on to say that he is not very good at this kind of mental fishing, but that what skill he does have he acquired not at school, but through . . . fishing: literally coarse fishing, with a rod and a float. When you are spending hours gazing at the red or yellow dot in the water in front of you, all the normal little nagging impulses that are competing for your attention gradually dissolve away, and you are left with the whole field of your awareness resting lightly but very attentively on the float, and on the invisible, autonomous world of water things suspended below it, and moving – perhaps – towards the surface, and towards your lure. Your imagination and your perception are both working on and in the water world. Thus fishing is an exercise which cultivates the kind of relaxed-yet-attentive, perceptive-yet-imaginative mode of mind that fosters intuition; and at the same time it offers a metaphor for the way in which such a mental attitude mediates between consciousness and the undermind.

This way of gathering and inspecting the fruits of intuition without bruising them, or avidly turning them into jams and pies, is, as Hughes says, something which people are differentially good at, or familiar with; and it is also an art which can be cultivated not just through literal fishing but through any form of contemplation that invites you to observe without interfering with the crepuscular world that lies between consciousness and the undermind; between light and dark; between waking and sleep. In the gloaming of the mind, if one is quiet and watchful, one can observe the precursors of conscious intelligence at play, and in so doing may be lucky enough to catch the gleam of an original or useful thought. As Emerson said in his essay on 'Self-reliance', talking of creativity: 'A man should learn to detect and watch that gleam of light which flashes across his mind from within . . . In every word of genius we recognise our own rejected thoughts; they come back to us with a certain alienated majesty.'[21] Several studies show that there are large differences between how well people are able to access these states of reverie, and that these correlate with how creative they are judged to be. People who have vivid imaginations, for example, being able to lose themselves at will in fantasy, or to recall childhood memories in great sensory detail, also score highly on standard tests of creativity.[22]

In similar vein, analytical psychologist James Hillman deplores the post-Freudian party game of 'interpreting' one's dreams. The dream, says Hillman, often has an integrity, an aura of both meaning and mystery, that is simply lost if one tries to dismember it into the familiar categories of thought. It is in the very nature of dreams to hint and allude. 'An image always seems more profound, more powerful and more beautiful than the comprehension of it.' To ask of a dream 'What does it mean?' is as misguided as to ask the same question of a painting or a poem – or of a sunset, come to that. 'To give a dream the meanings of the rational mind is . . . a kind of dredging up and hauling all the material from one side of the bridge to the other. It is an attitude of *wanting* from the unconscious, using it to gain information, power, energy, exploiting it for the sake of the ego: make it mine, make it mine.'[23] The proper attitude towards a dream, according to analytical psychology, is to 'befriend' it: 'to participate in it, to enter into its imagery and mood, to . . . play with, live with, carry and become familiar with – as one would do with a friend.' So 'the first thing in this non-interpretative approach to the dream is that we give time and patience to it, jumping to no conclusions, fixing it in no solutions . . . This kind of exploration meets the dream on its own imaginative ground and gives it a chance to reveal itself further.'

In some of the everyday problems we face, the 'goal' is clearly established in advance, and the value of the 'solution' has to be measured against predetermined criteria. If your car dies on the motorway, you want the emergency service person to get it going and fix the fault; you do not want them to start reupholstering the seats. But if a company's sales figures are declining, there is a whole range of possible 'goals' that one might pursue: advertising, customer service, market research, downsizing, product development, reorganising the structure . . . To have decided prematurely which aspect of the enterprise needs fixing may be to have missed a creative opportunity. The good intuiter is sometimes capable of delaying her decision about where she is going, even after she has set out. One of the areas where the value of this reluctance to specify the goal has been demonstrated most clearly is painting. Many artists have described the thrill of embarking on a canvas without knowing what will emerge. D. H. Lawrence, an enthusiastic amateur painter, described this vertiginous feeling.

> It is to me the most exciting moment – when you have a blank canvas and a big brush full of wet colour, and you plunge. It is just like diving into a pond – then you start frantically to swim. So far as I am concerned, it is like swimming in a baffling current and being rather frightened and very thrilled, gasping and striking out for all you're worth. The knowing eye watches sharp as a needle; but the picture comes clean out of instinct, intuition and sheer physical action. Once the instinct and intuition gets into the brush-tip, the picture happens, if it is to be a picture at all.[24]

A study of art students at the School of the Art Institute in Chicago by Getzels and Csikszentmihalyi looked in detail at the different working methods of the students as they tackled this task, and investigated whether there were any aspects of their modus operandi that correlated with the quality of the finished picture – as judged by the art tutors and practising artists. The students did indeed work in very different ways. From the large selection of objects available, from which to compose their still life, some students selected and handled as few as two, while others played with many more before settling on their selection. And some of them did 'play': they did not just pick the objects up; they stroked them, threw them in the air, smelled them, bit into them, moved their parts, held them up to the light and so on. The students also varied in the actual objects they selected. Some chose from the 'pool' those that were conventional, even clichéd, still-life subjects – a leather-bound book, a bunch of grapes. Others went for objects that were more surprising, or less hackneyed. Most interesting, though, were the differences in working practice once the students had started on their pictures. Some continued to change the composition of the objects, or even the objects themselves, for quite a long time, so that the finished structure of the picture did not emerge until rather late in the creative process. Others, once they had made their composition, stuck to it religiously, and their pictures took on recognisable form rather early.

The findings of the study were clear. The pictures produced by the students who had considered more objects, and more unusual ones, who played with them more, and who delayed foreclosing on the final form of the picture for as long as possible, changing their minds as they went along, were judged of greater originality and 'aesthetic value' than the others. What is more, when the students were followed up seven years later, of those who were still practising artists, the most successful were those who had adopted the more playful and patient modus operandi. These were clearly people who had learnt how to stay open to the promptings of their intuition, and who were comfortable setting out on a journey of discovery without the reassurance of knowing in advance where they were going. They are in good company. Picasso said of his own painting: 'The picture is not thought out and determined beforehand, rather while it is being made it follows the mobility of thought.'[25]

There is a whole variety of ways in which people differ with respect to intuition – and therefore an equal variety of ways in which we can set about trying to improve the hospitality of the conditions, both inner and outer, within which intuition can blossom. Being a 'mother of invention' is an art that we can learn. We can learn to acknowledge, and to take more seriously, the small seeds of poignancy and puzzlement that occur to us, and the gleams of thought that flash across the periphery of the mind's eye. We can discover the contexts and moods in which we are most creative and receptive, and make sure that we make time for these in our lives. We can guard against becoming too invested in a problem, and trying too hard. We can practise the art of not neatening problems up too quickly, and of not making up our minds too soon about what would count as a 'solution'. And we can cultivate patience. As the Tao Te Ching asks:

> Who can wait quietly while the mud settles?
> Who can remain still until the moment of action?
> Observers of the Tao do not seek fulfilment.
> Not seeking fulfilment they are not swayed by the desire for change.
> Empty yourself of everything.
> Let the mind rest at peace.
> The ten thousand things rise and fall
> While the Self watches their return.
> They grow and flourish and then return to the source.
> Returning to the source is stillness, which is the way of nature.

CHAPTER 6

Thinking Too Much? Reason and Intuition as Antagonists and Allies

> **The men of experiment are like the ant; they only collect and use. The reasoners resemble spiders, who make cob-webs out of their own substance. But the bee takes a middle course; it gathers its material from the flowers of the garden and of the field, but transforms and digests it by a power of its own.**
>
> *Francis Bacon*

Is it possible to think too much? Though people who cannot get to sleep for churning over their problems might answer in the affirmative, conventional wisdom sometimes seems to suggest not. In the classroom, the consulting room or the boardroom, we may operate as if the more analytical we were, the better. Or even if we do not, we may assume that we ought to; that a detailed listing and weighing up of considerations, for example, represents some kind of ideal cognitive strategy to which our actual behaviour approximates. As Benjamin Franklin wrote to the British scientist Joseph Priestley:

> My way [of tackling a difficult problem] is to divide half a sheet of paper by a line into two columns, writing over the one Pro, and over the other Con. Then during three or four days' consideration, I put down under the different headings short hints of the different motives, that at different times occur to me, for or against each measure . . . I find at length where the balance lies; and if, after a day or two of further consideration, nothing new that is of importance occurs on either side, I come to a determination accordingly . . . When each [reason] is thus considered, separately and comparatively, and the whole lies

> before me, I think I can judge better, and am less likely to make a rash step.[1]

The rationale for such a modus operandi is, presumably, that by making our thoughts and motives explicit and orderly, we can evaluate and integrate them better, and thus make better decisions. Or, as an influential textbook on decision-making puts it: 'The spirit of decision analysis is divide and conquer: decompose a complex problem into simpler problems, get your thinking straight in these simpler problems, paste these analyses together with a logical glue, and come out with a program for action for the complex problem.'[2] Be as explicit, as articulate and as systematic as you can be, and you will be thinking in the way that generates the best decisions and solutions. Given the evidence that we have looked at so far, however, we might have cause to question this ubiquitous, common-sensical assumption. D-mode and the slower ways of knowing work together, but they can get out of balance, and lose coordination.

Jonathan Schooler from the Learning Research and Development Center at the University of Pittsburgh has conducted a number of studies over the last few years that demonstrate graphically how thinking can get in the way of a whole variety of mental functions including everyday memory and decision-making, as well as intuition and insight. These studies go to the heart of the relationship between d-mode and intuition. One concerns how people choose between several possibilities, as when we are deciding which of a range of different foods we prefer. Schooler gave his subjects five different brands of strawberry jam to taste, and asked them to rate them and say which they liked best. The jams had recently been the subject of a consumer report, and those used had been ranked 1st, 11th, 24th, 32nd and 44th by the experts. Some of the subjects were told that they would be asked to explain the reasons for their choices, and to think carefully about their reactions and preferences. The results showed that those subjects who were left to their own devices ranked the jams in a way that corresponded closely to the judgement of the experts, while those who had been instructed to analyse their reactions disagreed with the experts.

Obviously this need not be a problem: maybe thinking about your choices makes you more independent. You decide what is right for you, rather than following the herd. If thinking carefully means that your decisions are based more closely on your true values and preferences, you would expect the thoughtful subjects to be *more* satisfied with their choices, and for this satisfaction to be more long-lasting. Unfortunately the reverse is the case. In a parallel study, subjects were shown five art posters and invited to choose one to take home. Those who had deliberated most carefully turned out to be significantly *less* satisfied with their choice, a few weeks later, than those who had chosen 'intuitively'. The deliberating jam tasters were more 'individual' than their intuitive colleagues, but they were worse, not better, at making choices that reflected what they *really* liked.

Neither of these choices, you could say, has a terribly important impact on people's lives. But deciding which courses to take in college certainly does. Schooler investigated university students as they selected their second-year psychology courses. They were given full information about all the options, including comments and ratings from those students who had taken each course the previous year, and asked to say which courses they thought they would take. Again, some students were asked to reflect in detail on the information provided, and the criteria they were using to make their choices. As in the other studies, those who thought most carefully were less likely to opt for the courses recommended by their peers, and more likely to change their minds subsequently. Later, when it came to signing up, the choices of the 'deliberating' students tended to revert to those that conformed more with received opinion, and correspond with the choices of those who had chosen intuitively.

On the basis of these studies, the researchers argue that there are a number of potentially negative effects of encouraging people to be more reflective and explicit about their decisions. In choosing a picture, or a jam, or a course, there are many interwoven considerations to be taken into account, *not all of which are (equally) verbalisable.* When the decision is made in an intuitive way, these considerations are treated in a more integrated fashion, and those that are hard to articulate are given due weight – which actually may be considerable. However, when people are forced (or encouraged) to be analytical, the problem is deconstructed into those considerations that are more amenable to being put into words. Thus the way the predicament is represented to consciousness may be, to a greater or lesser extent, a distortion of the way it is represented tacitly, and decisions based on this skewed impression are therefore less satisfactory.

In particular, d-mode may exclude or downgrade those nonverbal considerations that are primarily *sensory* or *affective*. Analytic thinking therefore tends to overestimate *cognitive* factors, which may be more easily expressed, resulting in decisions which seem 'sensible', but which fail to take into account non-cognitive factors. Additionally, the more carefully one analyses the different alternatives, the more one finds that there are good and bad aspects to each, and the greater the consequent tendency for judgements to become more moderate, more similar and therefore less decisive. Hence the tendency to come to decisions that differ from those which the acknowledged experts would have advised, and to feel obscurely dissatisfied with the choice one has made. Anyone who has ever agonised over a choice while shopping, and then regretted the decision immediately they have got the item home, will be familiar with this phenomenon. It is the dislocation between conscious and unconscious decision-making that people are referring to when they say that they should have listened to their 'heart', or their 'gut feeling', or their intuition.

As usual the issue is not black and white. We might now suspect that where a problem can be adequately represented verbally, and where the solution lies at the end of a logical chain of reasoning, a predominantly d-mode approach will be effective and efficient. While where the problem is more complex, contains aspects that are hard to articulate, or demands an insightful leap, d-mode will be less successful than a more receptive, patient approach. In another series of experiments, Schooler and his colleagues explored these two types of problems, looking particularly to see where active thinking helped and where it hindered.

'Insight problems' are those where people are in possession of all the information and ability necessary to solve them, but where there is a tendency to feel blocked or 'stumped', before suddenly having a kind of 'Aha!' experience in which the solution becomes immediately or rapidly obvious. In such problems, the difficulty is often caused by the tendency to make some unconscious assumptions that get in the way, or to fail to retrieve knowledge that would actually be helpful. We have met examples of such problems before – the 'mutilated chessboard', for example. Two different puzzles of this kind are shown in Figure 6.

The figure on the left represents a triangle made up of coins. The problem is to make the triangle point downwards by moving only three of the coins. The figure on the right represents a pen containing nine pigs. The problem is to build two more square enclosures that would put each pig in a pen by itself.

Contrast these with two so-called 'analytical' problems. In the first, imagine that there are three playing cards face down on the

Figure 6. Insight problems. (a) Move three coins to invert the triangle. (b) Draw two squares to give each pig its own enclosure. (The answers are in note 6.3 on page 236.)

table in front of you. You are given the following pieces of information:

> To the left of a queen there is a jack.
> To the left of a spade there is a diamond.
> To the right of a heart there is a king.
> To the right of a king there is a spade.

Your job is to say what the three cards are.

In the second problem, the police are convinced that one of Alan, Bob, Chris and Dave has committed a crime. Each of the suspects in turn has made a statement, but only one of the four is true. Alan said 'I didn't do it'. Bob said 'Alan is lying'. Chris said 'Bob is lying'. Dave said 'Bob did it'. Who is telling the truth; and who did the crime?[3]

In the two analytical problems, no additional knowledge has to be supplied by the problem-solver, and it is unlikely that any assumptions will be made unwittingly that would make the problems harder to solve than they already are. All that is required is a meticulous fitting together of the pieces of information – a non-trivial but in principle straightforward task – and the answer will emerge. We might imagine that, if people were asked to think out loud while they were attempting an analytical problem, their words might track their thoughts quite easily and accurately, and would be positively related to the actual solution.

But with the insight problems, we might argue that a different kind of 'thinking' is required, one which is more of the intuitive, behind-the-scenes kind; and in this case if people are required to think aloud, this might actually interfere with the intuitive process. As Schooler says in his paper: 'Verbalisation may cause such a ruckus in the "front" of one's mind that one is unable to attend to the new approaches that may be emerging in the "back" of one's mind.' Schooler's study did in fact analyse what subjects said as they were working on the two types of problem, and found exactly what had been predicted. Subjects who are solving analytical problems are neither helped nor hindered by the demand to think about and to verbalise what they are doing. But when they are attempting insight problems, they are very considerably hampered when they are simultaneously required to attend to and articulate what is going on in their minds.

In one variation of the experiment, subjects were informed before they started that they would be working on two kinds of problems, and that one kind, the 'insight' problems, would typically lead them into an approach that did not work. They were given an example of an insight problem, and told that if they got stuck, it would probably help to try to find a different approach or a new perspective. As before, some of the subjects were told to think aloud, and others were not. There were two interesting results. The first was that this heavy hint was of no benefit at all in solving the insight problems, and did nothing to offset the decrease in performance produced by thinking aloud. It looks strongly as if the way of knowing that leads to success in the insight problems is not only outside of conscious awareness, but outside of conscious control as well. If the processes of intuition are beyond voluntary control, then there is no way in which the subjects can make use of the 'helpful hints' which they have been given.

The researchers also found that the information about insight problems did have a marked effect on the solution of the analytical problems. When subjects were thinking aloud, the hint severely *damaged* their ability to solve the logical puzzles. Just as the attempt to solve insight puzzles through exclusive reliance on d-mode is misguided, so performance in analytical problems can be impeded if you sow doubt in people's minds about their straightforwardness. The suspicion that something might be trickier than it actually is causes the confident use of d-mode to falter, as people endeavour to seek out – intuitively – complexities that do not exist. This result reinforces the point that the selection of the right cognitive mode is a matter of appropriateness, and not of the absolute superiority of one way of knowing over another.

Listening to the tapes of people's verbalisations, it became clear to Schooler and his colleagues that the *contents* of the problem-solvers' thoughts, as they tackled the two types of problem, were different. People wrestling with the analytical problems talked fluently, and most of their comments referred to the problem itself. However, when they were attempting the insight problems, subjects paused more frequently, and the pauses were longer: there were many more occasions on which there was, seemingly, nothing going on in the problem-solvers' minds. And when people doing the insight problems did verbalise, they were four times more likely to make the kinds of comments that referred not to the logic of the problem but to their own mental state. They would say things like 'There is nothing that's going through my mind that's really in any kind of . . . that's in a verbal fashion'; or 'I know I'm supposed to keep talking but I don't know what I am thinking'. And this experience of 'nothing going on' was actually correlated with success on the insight problems. Those subjects who paused more solved more problems. Keeping up a running mental commentary really does interfere with the slower, less conscious processes going on at the back of the mind, and causes a drop in intelligence and creativity. We must presume that people for whom such chatter is habitual are thereby hampered when it comes to dealing with problems of greater subtlety or indeterminacy.

Jonathan Schooler's general point is of enormous significance. Some of what we know is readily rendered into words and propositions; and some of it is not. Some of our mental operations are available to consciousness; and some of them are not. When we think, consciously and articulately, therefore, we are not capturing accurately all that is going on in the mind. Rather we are selecting only that part of what we know which is capable of being verbalised; only those aspects of our cognition to which conscious awareness has access. We think what is *thinkable*; not what is 'true'. And the disposition to treat all problems as if they were d-mode problems thus skews our thoughts and our mental operations towards those that can be made explicit.

Other areas of our psychological life show similar effects. Take memory. There are many experiences that defy articulation, and our memories of them must therefore rely on non-verbal records. Our ability to recognise a huge collection of human faces, for example, with a remarkable degree of accuracy and effortlessness, attests to the power (and, I would say, 'intelligence') of these unspoken processes. What we can say about a face or an expression is a small fraction of what we can know. Thus it should come as no surprise to discover that the effort to *describe* a face so narrows our attention, and biases it towards the little that can be said, that memory is reduced quite severely. In another of his studies, Schooler gave subjects photographs of unfamiliar faces to study, and asked them to attempt to describe some of the faces but not others. These pictures were then mixed up with some new ones of rather similar-looking people, and the subjects were asked to pick out the ones they had seen before. The faces that had been described were recognised about half as well as those that had not, and this impairment was unrelated to how detailed or accurate a particular description had actually been. The same result is obtained if the stimuli are simple patches of colour.

The problem with description is twofold. First, the effort to describe the face forces one to break it up into its articulable features, and focuses attention on what can be said at the expense of what is genuinely (but non-verbally) distinctive. And secondly, at the time of recognition one may be trying to retrieve the 'written records' from memory, and match these to the pictures that one is being shown, rather than relying on the non-verbal, sensory records that have been registered. If this retrieval effect is a significant part of the problem, performance might be improved by preventing people from using the verbal 'code' while they are doing the recognition test. This could be achieved by forcing people to make their memory judgement very fast, perhaps. If you deprive them of the time it takes to *think*, they might have to fall back on the visual information which has been overshadowed by the verbal description, and thus overcome the interference. This is exactly what Schooler showed. When people had to make their recognition decisions quickly, the deleterious effect of verbalisation was removed.

In this case a 'snap judgement' is more reliable than a considered one. 'Decide first, and ask questions afterwards' may be the right strategy when dealing with non-verbal information. One can escape from the negative effects of d-mode by responding *faster* than thought, as well as more slowly.[4] The conventional wisdom that says we should always benefit from thinking and reflecting more is again seen to be in need of revision. There are interesting practical implications from this study for the handling of eye-witness testimony, identity parades, and so on. Asking witnesses to 'think carefully', and to describe what they have seen, may well interfere with their subsequent ability to recognise a face from a photograph or a line-up.

Schooler's studies have extended the range of everyday mental tasks with which articulation has been shown to interfere. In Chapter 3 we saw that learning to manage complex and unfamiliar situations, and to perform under pressure, can both be undermined by too strong a commitment to intellectual comprehension and control. Now we know that the same can be true when we are making choices and decisions, solving problems that involve insight, and even when we are simply recognising faces or other visual stimuli. However, again we should beware of falling into the trap of exaggerating the downside of d-mode. There is no value in demonising the intellect. There are many situations in life where an explicit grasp is useful or necessary. When we have to communicate our ideas to other people, in order to get practical tasks accomplished, we must obviously articulate as clearly as we can.

But we do not need d-mode just for communication. There are times when we need its analytical powers to test and refine ideas that have been thrown up by the undermind. The study of creativity in many different areas shows that what is required for optimal cognition is a fluid balance between modes of mind that are effortful, purposeful, detailed and explicit on the one hand, and those that are playful, patient and implicit on the other. We need to be able both to *generate* ideas, and also to *evaluate* them. Intuition is the primary mode of generation. D-mode is the primary mode of evaluation. Henri Poincaré summed it up when he said: 'It is by logic we prove; it is by intuition we discover.' For the scientist, intuition and contemplation may provide the vital creative insight that is both preceded and followed by the more disciplined procedures of d-mode. The chemist Kekulé, having first seen the cyclical form of the benzene carbon ring in a drowsy fantasy, concluded the report of his breakthrough to the Royal Society with the words 'Gentlemen, let us learn to dream. But before we publish our dreams, let us put them to the test of waking reason.' And Poincaré, having vaunted the need for patience, said: 'There is another remark to be made about the conditions of this unconscious work: it is possible, and of a certainty it is only fruitful, if it is on the one hand preceded and on the other followed by a period of conscious work.'[5] If we can think too much, it is also possible to think too little.

The classic formulation of scientific creativity, developed from Poincaré's observations by Graham Wallas in his 1926 book *The Art of Thought*, sees it as emerging from the interplay between four different mental modes or phases: preparation, incubation, illumination and verification. In the preparation phase, one gathers information, carries out experiments, and seeks, as hard as one can, for a satisfactory explanation – which obstinately refuses to come. D-mode is employed to the limit, and finally admits defeat. Then, as we saw in Chapter 5, the problem is put to one side to rest and incubate. If all goes well, at some unpredictable moment a new idea – novel, unexpected, but somehow full of promise – surfaces. And then, after this revelation or 'illumination', comes the return of d-mode, to apply its tests and checks, probing to see if the promise is fulfilled, and seeking ways to turn the illumination into a form which can be communicated, and which can compel the assent of others.

But it is not just scientists who value d-mode. Artists and poets too, though they are wary of its ability to strangle the creative impulse in its cradle, know that it is also a tool of which they have need. They are clear, as A. E. Housman said, that 'the intellect is not the fount of poetry, it may actually hinder its production, and it cannot even be trusted to recognise poetry when it is produced'. Yet many creative artists also speak of the value of a more deliberate, controlled, conscious mode of mind in sorting through the products of intuition, and shaping them into a finished product. Mozart distinguished between the conditions of creativity, and those of selection, when he said: 'When I am, as it were, completely myself, entirely alone, and of good cheer . . . it is on such occasions that my ideas flow best and most abundantly. Whence and how they come, I know not: nor can I force them. Those ideas that please me, I retain in memory . . .'[6] Not all the ideas that 'flow freely' are retained for future use: only those 'that please me'. John Dryden talks of intuition, or 'Fancy', 'moving the Sleeping Images of things towards the Light, there to be distinguish'd, and then either chosen or rejected by the Judgment'.[7] Even the Romantic poets such as William Wordsworth are equally clear that 'Poems to which any value can be attached were never produced . . . but by a man who, being possessed of more than usual organic sensibility, had also thought long and deeply. For our continued influxes of feeling are modified and directed by our thoughts.'[8]

Sculptor Henry Moore expressed in rather more detail the Janus-faced quality of the discriminating intellect.

> It is a mistake for a sculptor or a painter to speak or write very often about his job. It releases tension needed for his work. By trying to express his aims with rounded-off logical exactness, he can easily become a theorist whose actual work is only a caged-in exposition of concepts evolved in terms of logic and

> words. But though the nonlogical, instinctive, subconscious part of the mind must play its part in his work, he also has a conscious mind which is not inactive. The artist works with a concentration of his whole personality, and the conscious part of it resolves conflicts, organises memories, and prevents him from trying to walk in two directions at the same time.[9]

It seems as if full-blown creativity works in a way that is not unlike biological evolution. As long ago as 1946, R. W. Gerard suggested that imagination and intuition are to ideas what mutation is to animals: they create a diversity of new forms, many of which are less viable, less well suited to the demands of the environment, than those that existed already, but some of which, perhaps only a few, contain features and properties that are adaptive as well as novel. The undermind accounts for the 'arrival' of ideas, both fit and unfit. Reason and logic then act like the environment, putting each of these candidates to the test, and ensuring that it is only the fittest that survive.[10] (More recently neuroscientist Gerald Edelman has proposed, with his idea of 'neural Darwinism', that the development of different pathways in the brain itself is determined by a similar process. Those connections that 'work' to the animal's advantage are strengthened; those that do not fade away.)[11] The analogy is limited, however, by the fact that the undermind, unlike the process of genetic mutation, generates not just random variations of what exists already, but complex, well-worked-out candidates; not just guesses but *good* guesses, *educated* guesses. The undermind is intelligent in a way that mutation, as far as we know, is not.

Several artists talk of the need for conscious thought to come to the rescue of the creative process when intuition 'stalls', as it not infrequently does. If one is lucky, like Coleridge with 'Kubla Khan', the undermind does the whole thing for you. The creative product is 'channelled', and the only role left for the conscious mind is that of scribe. However, it is not always thus. Intuition will not be managed, and sometimes it seems to down tools before the job is completed. As Amy Lowell says: 'The subconscious is . . . a most temperamental ally. Often he will strike work at some critical point and not another word is to be got out of him. Here is where the conscious training of the poet comes in, for he must fill in what the subconscious has left . . . This is the reason that a poet must be both born and made. He must be born with a subconscious factory always working for him, or he can never be a poet at all, and he must have knowledge and talent enough to "putty" up his holes.'[12]

Housman, he who is most aware of the destructive power of the critical intellect, nevertheless had to draw on it in order to get a poem finished.

> Having drunk a pint of beer at luncheon . . . I would go out for a walk of two or three hours. As I went along, thinking of nothing in particular, only looking at things around me and following the progress of the seasons, there would flow into my mind, with sudden and unaccountable emotion, sometimes a line or two of verse, sometimes a whole stanza at once, accompanied, not preceded, by a vague notion of the poem which they were destined to form part of . . . When I got home I wrote them down, leaving gaps, and hoping that further inspiration might be forthcoming another day. Sometimes it was, if I took my walks in a receptive and expectant frame of mind; but sometimes the poem had to be taken in hand and completed by the brain, which was apt to be a matter of trouble and anxiety, involving trial and disappointment, and sometimes ending in failure.
>
> I happen to remember distinctly the genesis of the piece which stands last in my first volume. Two of the stanzas, I do not say which, came into my head, just as they are printed, while I was crossing the corner of Hampstead Heath between Spaniard's Inn and the footpath to Temple Fortune. A third stanza came with a little coaxing after tea. One more was needed, but it did not come: I had to turn to and compose it myself, and that was a laborious business. I wrote it thirteen times, and it was more than a twelvemonth before I got it right.[13]

The creative mind possesses a dynamic, integrated balance between deliberation and contemplation. It is able to swing flexibly between its focused, analytical, articulated mode of conscious thought, and its diffused, synthetic, shadowy mode of intuition. But the mind may lose its poise and get stuck in one mode or the other. And if its balance has been disturbed in this way, it takes time and effort to free it up again.

This process of rediscovering the complementarity of the mind's different modes has been graphically charted in a study of 'women's ways of knowing' by Mary Field Belenky and others.[14] They made a detailed study of the experiences of women of a wide variety of ages and backgrounds who were studying within the formal education system, and identified five stages through which these women seemed to pass on their journey towards expanded sophistication and confidence as 'knowers'. In the early stages of this development, they claim, many women, particularly those who have previously had little successful experience of formal education, start out feeling very powerless and inept in the face of the rational, articulate way of knowing. They feel as if they have no 'voice' of their own, and are in awe of others (principally but not exclusively men) whose d-mode voices sound loud, self-confident and authoritative.

But at some point, they may realise that they *do* 'know', and that there is validity to their experiences, feelings – and intuitions. In this stage of what the authors call 'subjective knowing', their respondents feel the first stirrings of their own 'epistemological authority'; though this is associated not with their ability to be rational and explicit, but with the emergence of a new respect for 'the inner voice'. 'Truth' is discovered not through argument and articulation, but through the promptings of gut feelings. It is as if there is 'some oracle within that stands opposed to the voices and the dictums of the outside world'.

> I just know. I try not to think about stuff because usually the decision is already made up inside you and then when the time comes, if you trust yourself, you just know the answer.

> There's a part of me I didn't even know I had until recently – instinct, intuition, whatever. It helps me and protects me. It's perceptive and astute. I just listen to the inside of me and I know what to do . . . I can only know with my gut. My gut is my best friend – the only thing in the world that won't let me down or lie to me or back away from me.

This discovery is experienced as vital and welcome; but it is accompanied, for some women, by an over-reaction in which thought-out knowledge is disdained as 'remote' and 'academic', while the inner voice is accepted as inevitably right and trustworthy simply by virtue of its 'innerness'. If it 'feels right', it is impossible for it to be wrong; even for it to be questioned can be taken as a mark of disrespect or felt as a violation. The sense of self as a valid knower is so precious, and yet so tenuous, that its source has to be defended against all conceivable threats, real or imagined. Instead of there being an absolute authority which is external, now this absolute is shifted inside. The feeling that there *is* an omniscient source of certainty remains; it is just relocated. In this move, the domains of logic, articulation and science may be completely rejected. The authors comment that: 'It was as if, by turning inward for answers, they had to deny strategies for knowing that they perceived as belonging to the masculine world.'

> It is not that these women have become familiar with logic and theory as tools for knowing and have chosen to reject them; they have only a vague and untested prejudice against a mode of thought that they sense is unfeminine and inhuman and may be detrimental to their capacity for feeling. This anti-rationalist attitude is primarily a characteristic of women during the period of subjectivism in which they value intuition as a safer and more fruitful approach to truth.

For some women this attitude may become as arrogant and offensive as that which they are at pains to denounce.

> A few of the women . . . were stubbornly committed to their view of things and unwilling to expose themselves to alternative conceptions. Although they might have described themselves as generous and caring, they could be, in fact, impatient and dismissive of other people's interpretations. They easily resorted to expletives when faced with others' viewpoints – 'That's bullshit!' . . . These were women at their most belligerent . . . adept at turning the tables on authorities by bludgeoning them with wordy, offensive arguments. In the classroom, as in life, they warded off others' words and influence via ploys to isolate, shout down, denigrate and undo the other.

When women look back on this stage later, from the more balanced, integrated perspective that Belenky refers to as 'procedural knowing',

> now they argue that intuitions may deceive; that gut reactions can be irresponsible and no one's gut feeling is infallible; that some truths are truer than others; that they can know things that they have never seen or touched; that truth is shared; and that expertise can be respected . . . They have learned that truth is not immediately accessible, that you cannot 'just know'. Things are not always what they seem to be. Truth lies hidden beneath the surface, and you must ferret it out. Knowing requires careful observation and analysis. You must 'really look' and 'listen hard'.

They have realised that their love affair with the inner intuitive voice – and particularly with the brittle certainty of its 'snap judgements'

– was a vital stage on the road to establishing their confidence in their own minds, and in developing their portfolios of ways of knowing; but that it was also tainted by the fear of uncertainty, and driven by more than a little wishful thinking. The 'inner voice' can easily be interpreted as telling you that things are the way you want them to be. As Minna, emerging slowly from a bad marriage and studying to become an occupational therapist, reflected: 'I was confused about everything. I was unrealistic about things. I was more in a fantasy world. You have to see things for what they are, not for what you want to see them. I don't want to live in a dream world [any more].'

At this later stage, knowing is characterised more by a respect for plurality and relativity, complexity and patience. The women in this study seem now to be discovering a more contemplative, less impulsive, form of intuition. The forced choice between feeling-laden subjectivity and remote objectivity begins to collapse, and knowing emerges from interaction and respect. It is no coincidence that, for these women, an interest in poetry often resurfaced at this stage. One of the women, a college senior, spoke scornfully of critics who use their 'so-called interpretations' as 'an excuse to get their own ideas off the ground'. She felt that to understand a text, you had to 'treat it as you would a friend'; accept it as 'real' and as 'independent of your existence', rather than 'using it for your own convenience and reinforcement'. She had become capable, in the words of Simone Weil, of a way of knowing that 'is first of all attentive. The soul empties itself of all its own contents in order to receive into itself the being that it is looking at, just as he is, in all his truth.'[15]

CHAPTER 7

Perception without Consciousness

At every moment there is in us an infinity of perceptions, unaccompanied by awareness or reflection; that is, of alterations in the soul itself, of which we are unaware because the impressions are either too minute or too numerous.

Leibniz

We are more in touch with, and more influenced by, the world around us than we know. In the 1960s, there was a belief that cinema audiences were being unconsciously manipulated into buying soft drinks that they didn't really want by messages flashed on to the screen too briefly to be detected consciously. Though it turns out that subliminal advertising is much less effective in persuading us to act against our best interests than we might fear, subliminal influences are indeed ubiquitous. They do not just occur when we are watching a screen or listening to a tape; they are present the whole time, and we could not manage without them. The undermind stays in continuous communication with the outside world without many of these conversations appearing in consciousness. Not only do we fail to *comprehend* what is going on in our own minds; we may not even *see* what is happening either.

Subliminal perception is hard to write about, because the relevant terms in the language are so muddled. It is symptomatic of our cultural neglect of the undermind that we have no word for 'being influenced by something of which we are unconscious'. Although it may conflict with the way some people use the word, I am going to use 'awareness' to refer to the general phenomenon of 'picking up signals' from the environment (or from the body), regardless of whether they get represented in consciousness. And I shall reserve 'consciousness', and '*conscious* awareness', for what appears before the mind's eye. Thus, in my usage, there is nothing paradoxical about using the expression 'unconscious awareness' to refer to the state of being affected or influenced by some stimulation that is not itself present in consciousness.[1]

In 1989, Thane Pittman and Robert Bornstein of Gettysburg College in Pennsylvania conducted an experiment to investigate how people decide between job applicants. Students were given the job specification – for a research assistant in the psychology department – and asked to review the applications of two young male candidates (call them Tom and Dick), and to make a recommendation as to which should be appointed. In fact, the applications differed in only one salient respect: Tom had good computing skills, but was poor at writing, while Dick was the opposite. Each application was accompanied by a photograph of a (different) young man. Before they took part in this study, the subjects had been asked to help out with another short experiment on visual perception, in the course of which they were exposed to five four-millisecond presentations of either Tom or Dick's face, accompanied by the word GOOD. An exposure of four milliseconds, under the lighting conditions used, is too short for any conscious impression to be gained. All the subjects saw was a tiny flicker of light, without any 'content'.

The students were twice as likely to choose, as the best person for the job, the candidate whose face had been subliminally presented. If it was Tom's face that had been projected, two-thirds of the students chose Tom over Dick. If it was Dick that had been flashed up, two-thirds opted for Dick. When they were asked to justify their decision, the subjects who had preferred Tom said it was because computing skills were more important for a research assistant than the ability to write well. The subjects who had chosen Dick said that anyone could learn the requisite computing skills, but to be able to communicate fluently was of vital importance.[2]

Pittman and Bornstein's subjects are unconsciously aware of the face that was flashed too briefly for conscious recognition, and of the association between that face and the positive evaluation conferred by the accompanying word GOOD. We know that they are unconsciously aware by virtue of the fact that their actual behaviour cannot be accounted for in any other way. Of course, it is true by definition that we cannot 'see' the unconscious itself directly. We can only infer its presence from its influence on things that we can observe, in just the same way that a 'black hole' cannot be photographed, yet its existence, and its properties, are implied by the fact that light behaves oddly in its vicinity. Likewise, to see how subliminal perception works, we have to look at the way people's actual behaviour, or their conscious thoughts, feelings and perceptions, are 'bent' by forces and processes that are themselves invisible.

Though there are many controlled studies of unconscious perception, we do not really need science to convince us of its ubiquity. We are constantly reacting to things that do not enter consciousness (though it takes an effort of will to notice the existence of things that one has *not* noticed). As you read these words your body is conforming to the chair or whatever you are sitting in or lying on, and you are adjusting your posture every so often in response to sensations of which you are usually unconscious. Your hands are responding to the size and stiffness of the book as you hold it and turn the pages. You may even, while reading, have the experience of realising that a clock is in the middle of striking the hour, and of being able to count up the number of unheard chimes that preceded your moment of 'awakening'.

The classic example of unconscious perception is driving a car and suddenly 'coming to'; you realise that you have driven for the last twenty minutes, absorbed in a conversation or a train of thought, without apparently – consciously – noticing anything at all of the road, the traffic or the operation of the controls. Consciousness has been absorbed in one world, while the unconscious 'automatic pilot' has been in quite another, coping very nicely on its own with roundabouts, traffic lights and pedestrian crossings. And the same ability to pursue flexible, intelligent routines, while being 'elsewhere', is manifest just as much in walking along a crowded street, doing the washing up, playing the piano, taking a shower or getting the children's tea. The ability to do things mindlessly, even cognitively quite demanding things like chatting to a friend or giving a lecture, is notorious. (You may recall the old story of the vicar who dreamt that he was delivering a sermon, and woke up to find that he was.)

The automatic pilot is sensitive to what is happening, and how things are going, just as the computer that controls the real automatic pilot in a plane is sensitive. On automatic pilot we do not just respond like stupid robots; we respond appropriately, like intelligent ones. Or we do most of the time. Sometimes we are caught out, and then we have behaved 'absent-mindedly'. You find that you are halfway to work before you remember that you were supposed to be going to the doctor's. Or, in William James' famous example, you go upstairs to change for the party, and suddenly realise that you are in your pyjamas and cleaning your teeth. Especially when 'we', that is, our conscious minds, are preoccupied, we may find ourselves pouring the hot water into the sugar jar or lighting the fire with today's newspaper.

But when consciousness is not so totally obsessed, merely entertaining itself with a fantasy or a rehearsal, then we do find that the unusual breaks through, it 'grabs our attention' and we wake up. A ball bounces out from between two parked cars just ahead, and suddenly we slow down and are on the alert for the child who may be about to dash out after it. Consciousness is re-engaged with perception and action; the conversation stops in mid-sentence. But here again there is evidence for unconscious perception, for how did we know to lock consciousness on to just *this* small detail, out of the stream of impressions that is constantly flooding in? How is it, when we impulsively turn our head, or stop and listen for a moment, that we frequently find there *is* something there to be attended to? The faint, unrecognised night noises of a friend's country cottage keep jerking me back from the edge of sleep; while the taxis that used to rattle past my London flat throughout the night left me unmoved.

The only possible explanation for these phenomena is that the undermind is keeping a continual check on what is happening below the horizon of conscious awareness, detecting what might be important or dangerous, and deciding when to butt in to consciousness with a 'news flash'. Of course it does this fallibly. Sometimes it interrupts me with false alarms – the noise which wakes me is just the beams creaking, not burglars or a fire – and sometimes it fails to alert me to what matters. But its existence is all I am trying to highlight at the moment, not its omniscience. The movements of consciousness, and the pictures it presents, reflect, like the news flash on the television, the judgements of editors, and the alertness of reporters whose existence we must infer, but whose faces we may never see.

We are constantly reacting to things not solely in terms of what they are, but in terms of what we expect them to be. We prepare ourselves, physically and mentally, for what is going to happen next on the basis of cues that frequently do not themselves enter consciousness. While this process continues in a routine and successful fashion, it usually remains unnoticed. But it reveals itself through its errors. The first few times you use a moving escalator, getting on and off feels slightly peculiar. But when you are more experienced, your body has learnt to make a subtle set of adjustments to help you keep your balance, which are automatically triggered by the surrounding visual cues as you step on to the stairs.

Now, as you approach, as travellers on the London Underground frequently do, an escalator that is stationary, the same cues trigger the same pattern, and you initiate, at just the right moment, a delicate compensation for an acceleration (or deceleration) that, disconcertingly, does not occur.[3]

An even more compelling demonstration of the same effect is provided by the room that you may be invited to enter if you visit the Psychology Department at the University of Edinburgh. The walls and ceiling of the room are actually an upside-down box that is suspended from cables just above the 'real' floor. The gap between walls and floor is too small to be noticed. You enter the room, the door closes behind you – and someone outside pushes the 'room' so that it moves relative to the floor, and to you. This produces a visual perception that normally only occurs when you yourself are walking or swaying, so, acting on this unconscious interpretation, you 'correct' your inferred movement by leaning in the opposite direction – and fall over.

Unconscious preparation may be perceptual as well as physical. Give someone a large and a small tin that weigh the same, and they will tell you that the small one is the heavier – because it is heavier than expected, given its size. Someone brings you what you unreflectingly imagine to be a cup of tea – and the first sip tastes strange, before you realise that it is actually a cup of coffee. Once your expectations are recalibrated, then its taste, the very same taste, becomes reclassified as familiar and satisfactory.

We do not see or taste or feel what is 'out there'; conscious perception is a useful fiction that misrepresents 'reality' in our own interests. As you read, your eyes are flicking along the lines of print in a succession of jumps and fixations – 'saccades' – yet what you see, consciously, is a whole, stable page of print. If you were to hold the book twice as far away from your eyes as it is now, it would hardly look any smaller, even though the image on your retina is only half the size. And though part of the page is (probably) falling on the 'blind spot' of the retina, you are not aware of a corresponding hole in the world you see. The undermind routinely makes all kinds of adjustments to the data it receives before it hands them on to consciousness – because it is usually advantageous to do so.

These everyday examples of unconscious perception often concern aspects of the world which are perfectly visible, audible and so on, but which, though they are registered by the undermind, do not make it into consciousness. They affect us, but pass unnoticed. Experimental studies such as Pittman and Bornstein's, however, have tended to use stimuli that are themselves very faint or fleeting, on the borders of what is perceptible. Such situations demonstrate very clearly the nature of unconscious perception. And they have fascinated psychologists since the very beginnings of the scientific approach to the study of the mind.

In a classic study in 1898, for example, B. Sidis showed people cards on which were printed a single number or letter – but the cards were placed far enough away that his subjects were quite unable to read what was on them. Sidis reports that 'the subjects often complained that they could not see anything at all; that even the black, blurred, dim spot often disappeared from their field of view'. However, when he asked them to name the characters on the cards, they were correct much more often than they would have been by pure guessing, even though that is exactly what they felt they were doing. Sidis concluded from his experiments that there is 'within us a secondary subwaking self that perceives things which the primary waking self is unable to get at'.[4]

Even earlier, in 1884, philosopher C. S. Pierce carried out a series of tests with his graduate student Joseph Jastrow at Johns Hopkins University in America, in which they judged over and over which of two nearly identical weights was in fact the heavier. Again, despite the fact that their subjective confidence was effectively zero, they were able to do much better than chance would dictate. Over the thousands of trials on which they made complete guesses, indicating 'the absence of any preference for one answer over its opposite, so that it seemed nonsensical to answer at all', they were in fact correct between 60 and 70 per cent of the time. What was particularly interesting about their study was their recognition that these results were not just of curiosity value; they are of real significance for the way people operate in the world, for instance how we relate to each other. They wrote that:

> The general fact has highly important practical bearings, since it gives new reason for believing that we gather what is passing in one another's minds in large measure from sensations so faint that we are not fairly aware of having them, and can give no account of how we reach our conclusions about such matters. The insight of females as well as certain 'telepathic' phenomena may be explained in this way. Such faint sensations ought to be fully studied by the psychologist *and assiduously cultivated by everyman.*[5] (Emphasis added)

Not only do they see the relevance of these findings for everyday life; they also suggest that people may increase their sensitivity to such faint sensations. Just as intuition can be educated and sharpened (as I argued in Chapter 5), so can one's ability to make use of the mass of weak impressions that underlie – and are usually neglected by – our normal ways of seeing and knowing. It is as if the mind has two thresholds, one below which it registers nothing at all, and a second above which something becomes conscious. In between the two lies the demi-monde of the undermind in which impressions are active but unconscious. And Pierce and Jastrow's suggestion is tantamount to saying that the distance apart of these two thresholds can vary, so that it is possible to increase one's *conscious* sensitivity to what had previously been going on at an unconscious level.

An early demonstration of the power of unconscious perception to influence what does appear in consciousness was given by Otto Poetzl, a Viennese neuropsychologist working with casualties of the First World War. He tested a number of soldiers who had suffered gunshot wounds to the part of the brain which processes visual stimuli, the occipital lobe at the back of each cerebral hemisphere, and discovered something rather odd. They were effectively blind in the centre of their vision, yet if they kept their eyes fixed on a picture (which they couldn't 'see'), ideas and images would begin to come into their conscious minds which were clearly related to the 'invisible' picture. Associations to the picture would start to surface in consciousness not as features of a coherent visual perception, but as more free-floating and mysterious fragments. Poetzl wondered whether he could produce the same kind of effect in people with normal vision, and devised the following test. First a detailed picture was flashed very briefly, for just one hundredth of a second, in front of the volunteers. They were asked to draw whatever they could of what they had seen – which was usually nothing or very little. Then they were told (absurd though it may sound) to go away and to have a dream that night, and to come back the next day, relate their dreams, and draw any elements of their dreams that they could. When he analysed the records of the dreams, he found that they contained many fragments and associations of the original 'invisible' picture.[6]

Though these early studies were not as tightly designed as one might wish, the essential results have been replicated recently under more stringent conditions. Cambridge psychologist Mark Price, for example, has shown that people are able to 'guess' the category to which a word belongs, even though the word itself was exposed too quickly to be detected. If subjects are flashed the word 'carrot' subliminally, they may not be able to say what the word was, but they still have a better than chance likelihood of guessing that it was a vegetable. In one of his experiments, when the subject happened to be his brother, Price inadvertently replicated the Poetzl effect. At one point he flashed the word 'camel' which his brother failed to detect. However, in the middle of the following presentation, the subject suddenly started chuckling to himself. When Mark asked him what was up, he replied that he was laughing because an absurd fantasy about camels had suddenly popped into his head 'from nowhere'.

Similar effects are obtained when stimuli are hard to detect, not because they are faint or fleeting, but because they occur in peripheral rather than central vision. John Bradshaw at Monash University in Australia has shown that we can unconsciously read words that are on the edge of the visual field, and that such subliminal perception will influence how we interpret what we are consciously attending to. He flashed people an ambiguous word like 'bank' in the middle of the screen, and simultaneously, at the edges of vision, he flashed another word, such as 'money', which suggested one of the two different meanings of the central word. Though they had no conscious awareness of the peripheral words, his subjects nevertheless tended to interpret the central word as meaning 'financial institution' rather than 'edge of a river'.[7] Information which is strong enough to exceed the threshold of awareness, but not to become conscious itself, is nevertheless able to influence what *does* appear in consciousness. As we saw with intuition, consciousness seems to demand evidence that is more definitive than does the undermind.

Another implication of the Pittman and Bornstein study discussed earlier is that, because subjects do not know *how* or even *that* they have been influenced subliminally, they are unconscious of the true source of their decision, and they are therefore unable to use consciousness to guard against the influence itself. They are susceptible to the subliminal message precisely because they do not know it is there. People may have many tendencies that they try to control or mitigate, but unless these are picked up by the 'radar' of consciousness, those controls may not be able to operate. Just as we saw with intuition, the undermind may work with a richer database than consciousness, but the tapestry it weaves may contain threads and assumptions that are false or out of date. This opens up the intriguing possibility that, by bypassing the checks and inhibitions of consciousness, a subliminal stimulus may actually have a greater effect on behaviour than one perceived clearly.

A study by C. J. Patton gives a graphic demonstration of this effect. She chose as her experimental subjects two groups of female college students, one comprising women with 'normal' attitudes to eating, and the other women who had a history of eating disorders. Clinical data suggested that many women in the latter group would have developed a very ambivalent attitude to food, on the one hand craving it as a source of comfort, and on the other disliking themselves for being so dependent, or for becoming overweight (at least in their own eyes) as a result, and struggling to control the desire to eat, especially when in company. The question was: how would these women behave if they became anxious, but were unaware of either the fact or the source of their anxiety?

Like the Pittman and Bornstein study, Patton's experiment involved two parts. In the first, all the women took what was described as a 'visual discrimination test', in which they were to try to identify sentences that were flashed on a screen. There were two key sentences: one, which has previously been shown to be capable of inducing some anxiety in many subjects, was 'Mummy is leaving me', and the other, a neutral control, was 'Mummy is loaning it'. The sentences were flashed either subliminally, for four milliseconds, or consciously, for 200 milliseconds. Half of each of the two groups was flashed one of these messages, and the other half the other. Following this, the subjects went on to take part in the second experiment, in which they were to do a taste test on three different types of biscuit. After explaining this latter task, the experimenter left each subject alone with three full bowls of biscuits to complete the test. After the supposed taste test was completed, the experimenter checked to see how many biscuits each subject had eaten.

The results showed that the neutral sentence had no effect on biscuit consumption for any of the subjects, regardless of whether it was consciously seen or not. However, after having been shown 'Mummy is leaving me', the women with the eating disorders consumed twice as many biscuits as the other women – an average of twenty – *but only when the sentence had been projected subliminally.* For the 'normal' subjects at both exposure durations, and for the eating-disordered subjects when the sentence was clearly visible, the number of biscuits eaten was the same as in the neutral, control condition. When perception is conscious, it becomes possible to override and control our impulses. When we are unconscious of having been affected, we are less able to be vigilant.[8]

A similar effect can be produced if our attention is distracted from the crucial information. J. M. Darley and P. H. Gross asked people to estimate the intellectual ability of some hypothetical children, given various kinds of information. When the only background they were given was the parents' jobs and incomes – their 'socio-economic status' – this information did not influence their judgements. However, another group of subjects were additionally shown a videotape of the children apparently performing somewhat ambiguously on an intelligence test; that is, doing well on some of the test items, and poorly on others. Under these circumstances, when people had been told that one of the children they had watched came from a poor family, they would rate that child's intellectual ability lower than that of an equivalent child whom they had been led to believe came from a better-off family. Apparently, when people are aware that they *might* be biased by such information, they can take steps to compensate. But the effort of focusing on interpreting the child's concrete performance on the test seemed to mask the subjects' awareness of the need to be vigilant about their own stereotypes, and thus their assumptions were able to sidle into their judgements unnoticed.[9]

An experiment by Larry Jacoby at McMaster University in Canada emphasises both the depth of the unconscious interpretations we make, and the power that conscious awareness has to reorganise these interpretations. His study shows that the unconscious can even mislead us about what *kind* of experience we are having: whether it is a new perception or a memory, for example. These basic categories of experience are not 'given'; they, too, are judgements or attributions about which it is possible to be mistaken. Jacoby has shown that these judgements tend to be heavily influenced simply by how easily or fluently something is processed by the undermind. We know from many previous studies that having recognised something once makes it easier to recognise again; there is a residual effect of the first recognition that facilitates the second. And it may be this relative ease of processing that underlies the judgement that something *has* occurred in the recent past. The decision to treat an experience as a memory, rather than as a fresh perception, is, at least in part, an inference based on the fact that we were able to identify and categorise it faster than might have been expected. If this is so, we might be able to trick the undermind into calling something a memory by making it easier to process. (There is clearly the basis of a psychological explanation for *déjà vu* experiences here.)

Jacoby's experiment managed to create exactly this confusion.[10] He and his colleagues showed their volunteers a list of words, and after a short delay, showed them another longer list, one word at a time, that contained the words from the first list mixed up with some new words. Subjects were asked to say, of each word on the second list, whether it had been on the first list or not: in other words, to classify their experience of each word on the second list as a 'memory' or just a 'perception'. The experimenters' cunning manipulation was to make some of the test words slightly easier to read than others by varying the clarity of the print. They found that new words in clearer type were likely to be falsely 'recognised' as having been on the original list. Unconsciously people thought that the relative ease of processing the clear words was due to having seen them before, so they judged them to be memories.

Recent laboratory studies of 'false memory syndrome' have shown that judging an experience to be a 'memory' (and therefore 'real'), rather than a fantasy, is also influenced by the nature of the experience. It may be that the more vividly you can get someone to fantasise, the more likely they are subsequently to misrepresent this experience as a true memory.[11] Certainly, it is not uncommon to experience some confusion, especially just after waking, as to whether one is recalling a lifelike dream from the night before, or a memory of an event several days ago.

There is an additional feature of Jacoby's study that is very significant. The illusion of memory is removed if people are told (or if they notice for themselves) that the visual clarity of the second batch of words is being manipulated. If they are conscious that ease of processing is being influenced by another factor, they can take this into account in their judgements. They can keep the two variables separate. But if people don't know what is going on, then the undermind bundles the two different *sources* of ease – memory and clarity – together, and comes up with some wrong answers. As with the previous examples, when we are consciously aware of an influence, we may be able to guard against it or compensate for it. We can see that what we are doing *is* making an inference or an assumption. When we are not aware of the same influence, when it comes to us subliminally and is, so to speak, already dissolved in our perception by the time it has arrived in consciousness, then we implicitly *trust* it.

In everyday life, this phenomenon of self-monitoring has a large impact on how we form, and deal with, our stereotypes. For

Figure 7. What does the doctor reply?

example, look at Figure 7 and ask yourself what the doctor might reply.

Many people think that the doctor might say something like 'You look quite slim to me', while thinking to himself 'I wonder if we have a case of anorexia nervosa here?' In fact what the doctor says is: 'Don't worry; a lot of men tend to put on weight around your age.' Even those who pride themselves on their sensitivity to gender issues may unconsciously be trapped by the picture into assuming that the man is the doctor. When the stereotype is unconsciously stirred into the perception, then we may start to try to make sense out of a seemingly puzzling situation – not realising that it is the assumption that is problematic, not the reality. But of course once we are *aware* of the assumption, the whole picture – literally – changes.[12]

There is a third feature of unconscious perception which the original experiment by Pittman and Bornstein reveals, and that is the tendency for consciousness to 'fill the explanatory gap' with a plausible story, and not to recognise that this is what it has done. Their subjects 'explained' their selection of one job candidate over the other on the basis of an apparently rational appraisal of the relative importance of computing and writing ability. They did not offer this as a *conjecture* about their thought processes, but as a bona fide account of what actually happened in their minds. Yet the evidence shows that they are mistaken. Their choice is clearly and powerfully influenced by the manipulation of which they have no conscious knowledge. Their explanation of the choice, though they genuinely believe it to be true, is actually based on what would have been plausible. They do not intrinsically value computing over writing (or vice versa); they are just trying to generate a rationale for how their preference came about.

This tendency to confabulate is not an isolated or occasional phenomenon. There is plenty of evidence that we do it much more than we think. Occasionally we may acknowledge that there is an element of surmise in our reasoning, as when we account for behaving unreasonably by saying 'I *must have been* tired'; but often we buy our own reasoning uncritically, and with complete conviction (and confidently assert that 'I snapped *because I was* tired'). There are now many experimental illustrations of the ways in which we misconstrue our own motivation. In one, a 'street trader' laid out several different pairs of tights and invited people to say which they preferred. Whichever order they were arranged in, it turned out to be the tights that were on the right-hand end of the row that were chosen most frequently. Clearly there is a statistical bias towards choosing the item in that position, regardless (within limits, obviously) of what it is. Yet when asked why they had chosen the tights they did, no one said 'because they were at the right-hand end'.[13]

In another series of studies on the so-called 'bystander' effect, people were observed in a real-life situation, waiting for a train for example, and a stooge on the platform would suddenly fall to the ground and start groaning. The question is: who goes to the person's aid? What is observed, across a variety of different conditions, is that the more people there are around, the less likely any one person is to offer help. But if you ask those who stood by *why* they didn't get involved, they will tell you all sorts of stories which make no mention of the number of other people. And if you suggest to them that the number of bystanders might have influenced them, they will dismiss the possibility out of hand. In all these cases, people do not know they are confabulating – and would be most indignant if you were to suggest they were. Their conscious interpretation *is* their reality.[14]

The more we acknowledge the existence of the undermind, and its incredible ability to register events and make connections, the less we may need to turn to magical explanations for mental phenomena that at first sight appear strange or supernatural. Take what is sometimes referred to as the 'sixth sense', the rather mysterious ability which is sometimes invoked to account for the experience of somehow 'knowing' that you are being looked at, or that there is someone else in a room that you had supposed to be empty. But is it a sixth sense that is at work, or merely a collection of unconscious impressions derived from the other five? Perhaps this form of intuition could be explained on the basis of a collection of minimal unconscious impressions derived from the five ordinary senses, each of which is too weak to impinge on consciousness itself, but which nevertheless add up to an inexplicable 'feeling'? There do not seem to be any empirical studies of this, but the possibility is effectively described in *Tender is the Night* by Scott Fitzgerald, who was himself fascinated by the activities of the cognitive unconscious.

> In an inhabited room there are refracting objects only half noticed: varnished wood, more or less polished brass, silver and ivory, and beyond these a thousand conveyors of light and shadow so mild that one scarcely thinks of them as that, the tops of picture-frames, the edges of pencils or ash-trays, of crystal or china ornaments; the totality of this refraction appeal-

> ing to equally subtle reflexes of the vision as well as to those associational fragments in the subconscious that we seem to hang on to, as the glass-fitter keeps the irregularly shaped pieces that may do, some time. This fact might account for what Rosemary afterwards mystically described as realizing there was some one in the room, before she could determine it.[15]

We might even venture that a heightened subliminal sensitivity to other people, or even to the contents of one's own mind, might account for some 'telepathic' phenomena. As Pierce and Jastrow suggested, 'we gather what is passing in one another's minds in large measure from sensations so faint that we are not aware of having them . . . certain telepathic phenomena may be explained in this way'. It is possible that their thinking on the matter may even have been influenced directly by a study, also reported in the 1880s by the French physician, philosopher and psychologist Theodore Flournoy, of the renowned medium Catherine Muller, or 'Helen Smith' as she was pseudonymously known. In her meetings she would fall into a trance and undergo personality changes in which she would re-enact scenes from her previous lives. She became, in turn, a fifteenth-century Indian princess, Marie Antoinette, and a visitor from the planet Mars, in which latter incarnation she was able to talk in 'Martian', and discuss the planet's landscape, vegetation and people in extraordinary detail. All her 'characters' were most convincing, and their 'messages' for her clients often highly pertinent and perceptive. Flournoy gained her trust and searched open-mindedly for a 'natural explanation'; one which would credit her with being neither a genuine space and time traveller, nor a fraud.

By meticulously researching her early life, Flournoy was able to show – just as Lowe had done for Coleridge's 'The Ancient Mariner' – that much of her material had come from books she had read as a child and had, consciously, completely forgotten. He described her behaviour under trance as 'romances of the subliminal imagination', and each character represented a reversion of her personality to a different phase of her childhood. His analysis of 'Martian', he argued, showed that it was based on the syntax of French, although the linguist Victor Henry, who also studied her, contended that much of the Martian vocabulary was derived from Hungarian – the mother tongue of Helen Smith's father.[16] In this, as in other cases, Flournoy found no evidence to convince him that her perceptions of other people could not similarly be accounted for on the basis of some combination of buried knowledge and an acute subliminal sensitivity to non-verbal, so-called 'paralinguistic' cues.

No such investigation, however scrupulous, demonstrates that genuine reversion to past lives, or contact with the spirit realm, does *not* occur. And any particular case is always arguable. But such careful and even-handed investigations do at least require us to respect the powers of the undermind, and they might, regrettably to some people, advise caution in interpreting such exotic phenomena as out-of-the-body and near-death experiences, clairvoyance and divination, and so on. There are those who see such experiences as incontrovertible evidence of supernatural powers or influences, and use them as 'proof' that there is more to life than is dreamt of in our current psychology. It is often claimed that they could not have come from 'within'; that there must be real spirits out there who are talking to us, or real powers of telepathy which defy the known laws of physics and physiology. Maybe there are. But the case for the very existence of some of these strange experiences, let alone any particular explanation, is not yet established. And at least in some cases the magical conclusion may be premature – because the role of the *unconscious* has not been fully appreciated. The implicit identification of mind with consciousness is manifested time and again in esoteric circles in the assumption that, if we cannot find an explanation for some phenomenon in conscious terms, then 'the mind' cannot have been responsible at all.

CHAPTER 8

Self-Consciousness

Ah, what a dusty answer gets the soul
When hot for certainties in this our life
George Meredith

During the course of the second year of his therapy, a middle-aged client, a man of considerable intellect and accomplishment, was talking about the negative patterns in his life. The therapist, Joseph Masling, observed: 'You seem to think you have no right to be happy.' The man immediately began to fidget almost uncontrollably, before eventually subsiding into stillness. After a long silence, he said, 'What did you say?' Another of Masling's clients, a young woman on the verge of successfully completing her graduate training programme, behaved in almost exactly the same way when Masling commented, 'Have you noticed how much easier it is for you to tell me about your failures than your successes?' She too, after a lot of squirming, had to ask him to repeat what he had said.[1]

The undermind is a layer of activity within the human psyche that is richer and more subtle than consciousness. It can register and respond to events which, for one reason or another, do not become conscious. We have at our disposal a shimmering database full of pre-conceptual information, much of which is turned down by consciousness as being too contentious or unreliable. Conscious awareness decides what it will accept as valid – and thereby misses dissonant patterns and subtler nuances. While in d-mode, consciousness tends to present to us a world that is somewhat cautious and conventional. Sometimes this is appropriate, but if we get stuck there and lose the key to the twilight world that subserves it, we mothball valuable ways of knowing which can find sense and weave meaning out of a collection of the faintest threads and scraps.

I have suggested that one way of expressing this disparity between conscious and unconscious is in terms of two thresholds, a lower one, above which the undermind becomes active, and a higher one, above which information enters consciousness. The closer together these two points are, the more 'in touch' with the unconscious we are, and the more complete is our conscious awareness of what is happening across all the mental realms. The further apart they are, the more our conscious perception is impoverished. This quantitative notion of thresholds is rather crude, but it enables us to formulate an important question: what is it that determines how near or how far apart the two thresholds are? More generally, is the relationship between conscious and unconscious forms of awareness a dynamic one, subject to change, and if so, what are the forces that control it? In cases such as Masling's, it seems clear that some information received unconsciously can cause a considerable amount of non-verbal discomfort, and that, as a result, it is gated out of consciousness. The therapist's remark occasions a very rapid raising of the conscious threshold. Perhaps it is specifically things that are threatening that cause the conscious threshold to shoot up.

Corroboration for this supposition comes from a phenomenon called 'perceptual defence', which has been known to experimental psychologists since the 1940s. In the classic version of these studies, a subject is repeatedly flashed a word very briefly, and the exposure duration is gradually increased until she is able to identify the word correctly. Some of the words used are neutral, while others are vulgar or disturbing in some way. The charged words do not become consciously visible to the subjects until they are exposed for a considerably longer duration than the neutral words. If recognition and consciousness are the same thing, this result is simply incomprehensible. How could one selectively raise the perceptual threshold for things that have not yet been recognised? Unconscious perception provides the only explanation: the taboo word *is* recognised unconsciously, and the upper, conscious threshold is immediately raised in order to try to protect consciousness from the threat or emotional discomfort that the word has generated.[2] Jerome Bruner, one of the instigators of research on unconscious perception back in the late 1940s, used to use the analogy of the 'Judas eye', the peephole used by the doorkeeper at a 'speakeasy' to distinguish between bona fide members, for whom the door opens, and undesirables, such as the police, who are shut out. Without the Judas eye, one could only tell friend from foe by opening the door – and then it was too late.

Conversely, we can demonstrate that access to information in the undermind that is dubious, not because it is directly threatening, but because it is faint or ephemeral, can be increased by making subjects feel more relaxed and 'safe'. One way to do this is to ask subjects to express this weak information without feeling that they are under pressure, or being judged in any way. Normally when people are asked to recall something previously shown, they feel that *they* are being tested.[3] Psychologists' experiments are designed to be hard enough for people to make some errors: if everybody got everything right, the data would not differentiate between different conditions, and it is these differences that tell us interesting things about how the mind is working. And nobody likes making mistakes. It may be that the normal type of memory test, for example, underestimates how much people really do know, because the feeling of being on trial makes them adopt a cautious attitude.

Experiments by Kunst-Wilson and Zajonc and others have managed to demonstrate this effect. First they show people a sequence of complicated nonsense hieroglyphics. When these are subsequently mixed up with some new shapes of the same kind, subjects are rather poor at picking out the ones they saw before. However, if, instead of asking people in the second part of the experiment to *recognise* the 'old' squiggles, they are simply asked to point out the ones they *prefer*, they tend quite reliably to choose the ones they saw before. When self-esteem is at stake, delicate unconscious forms of information and intelligence seem to be disabled or dismissed, and the way we act becomes clumsy and coarse. When we are less 'on our best behaviour', the glimmerings of knowledge from the undermind are more available to guide perception and action. Sometimes reception is good, and we are able to pick up and use the undermind's faintest broadcasts. At other times, when we are stressed, only the strongest stations get through.[4]

The same kind of release from pressure can be achieved by presenting the 'test' as if it were a guessing game, rather than a measure of achievement. When we treat something as a 'pure guess', we do not feel responsible for it in the same way. We are freed to utter things that come to us 'out of the blue', because there is no apparent standard of correctness or success against which they, or we, will be judged. One method that has been used to investigate this idea was first devised to assess the memories of people with severe retrograde amnesia. Such people seem to have completely lost their ability to remember what has happened more than a few seconds ago. If you meet one of these people, then go out of the room and come back five minutes later, they will greet you as if you were a stranger. Give them a short list of words to study, take the list away, and after a short delay ask them to recall the words, and they will look at you blankly and say 'What words?' Yet there has for many years been a suspicion that these patients *do* have some memory; it is just that they are unable to access it *deliberately*.

The nineteenth-century French physician Claparede, for example, concealed a pin between his fingers when he was introduced to one of these amnesic patients, giving him a prick as they shook hands. On leaving the room and reappearing a few minutes later, Claparede was, as expected, treated by the patient as if they had never met before – yet the patient was curiously reluctant to shake his hand. When queried about this antisocial behaviour, he rather vaguely explained that 'you never know with doctors; sometimes they play tricks on you'.[5] It is no coincidence that it was a *painful* stimulus that was unconsciously registered, for the undermind is particularly concerned with things that are of significance for our survival and wellbeing.

The suspicion that amnesiacs have more memory than it appears has recently been confirmed in the following way. Subjects are given some time to study a list of words which they are asked to remember, and then the list is removed. A little later, instead of asking the subjects to recognise or recall the words in the conventional manner, they are shown the first two or three letters of a word and asked to respond with the first word beginning with those letters that comes into their heads. As far as the patients are concerned, this a completely new exercise. But the prompts they are given have been selected so that they can be completed with one of the words on the original list – and this to-be-remembered word has been chosen to be less common in the language than some alternative words that could also be used to complete the frame. So if one of the words on the original list was CLEAT, the patients are asked to think of a word that completes the frame CLE– –. Without any memory of the list, people would respond to the cue with a more common word like CLEAN or CLEAR, but in fact the amnesic patients tend to produce the rarer word which they recently saw, but cannot 'remember'.

The words must have been recorded, but the memories only reveal themselves in the way they influence the (apparently) 'free' association. It begins to look as if the 'amnesia', in these cases at least, is more to do with an inaccessibility of memories to consciousness than with an inability to register what has happened. This same effect has now been reproduced in people with undamaged memories, using subliminal perception techniques. Subjects are presented with a number of words on a screen, one by one, too briefly for conscious perception. If they are subsequently asked to recall the words, they will, like the amnesiacs, say 'What words?' Yet, if they are tested with the same 'free association' game, the words they did not even 'see' are found to have made a significant difference to how their minds are spontaneously working. They behave exactly like the memory-impaired patients.[6]

The same kind of distrust of faint information by consciousness has recently been demonstrated by Cambridge psychologist Tony Marcel in a neat study that focused on perception rather than memory.[7] His (unimpaired) subjects were presented with very weak flashes of light, so weak that it was hard to tell whether there was anything there or not, and asked to indicate each time they thought they saw a flash. But Marcel asked them to indicate in any of three different ways: by blinking their eyes; by pressing a button; or by saying 'Yes [I see the light]'. He discovered that these three different methods of answering the same question were not equivalent but gave quite varied results. When people were blinking, they 'saw' many more of the weak flashes than when they were replying verbally, with the button-pressing somewhere in between. When the subjects were asked to respond to the light by *both* blinking and reporting, there were many occasions on which the eyes said Yes while the voice said No. (By measuring the time intervals between stimulus and response, Marcel was able to exclude the possibility that these results could have occurred simply because of reflex blinking.)

Marcel points out how these results challenge our commonsense view of the mind. Our normal assumption is that we have a unitary consciousness which registers 'what's there'. If something is 'there', you can sense it and report it, and the *way* you report it ought not to make any difference to whether you 'see' it or not. Responding is, we assume, 'downstream' of perception, and what comes later in the processing chain should not affect what comes earlier. Under ordinary conditions, this assumption seems to work. But with stimuli that are ambiguous as to whether they are 'there' or not, that model of ourselves begins to break down. The method of report now appears to have a retrospective effect on what we see. The medium that is most closely tied to normal consciousness – verbalising – turns out to be the one that 'sees' least well, while the one that is most automatic, most engrained, most *un*conscious, turns out to be the most sensitive of all. There is corroboration for the idea that *the more the self is involved, the more cautious consciousness has to be,* for fear of 'getting it wrong'. Clearly a verbal report, 'Yes, there was a flash', or 'No, there wasn't', feels like more of a personal commitment than the mere blink of an eyelid – an act that one does not usually think of as requiring close personal supervision or involving much ego investment.

There was another aspect to the experiment which directly replicated the beneficial effect of 'guessing', and its superiority over 'trying'. In each of Tony Marcel's studies, regardless of which modes of response were being used, subjects' ability to detect the weak light was less than perfect. However, when Marcel asked people not to try to report the presence of the flash accurately, but simply to guess, their performance magically shot up to almost 100 per cent! To 'try' is to have some kind of investment in the outcome. You *care*, you *bother* – and therefore you cannot help but be 'bothered' if your effort proves unsuccessful. With a pure guess, on the other hand, you feel as if you are plucking an answer out of thin air. And, as we have seen before, when the pressure is reduced, you are able to allow your choice to be guided by unconscious promptings that *are* adequate for the task, despite consciousness's lack of faith.

Twenty-five years ago, when I was beginning my graduate work in psychology at the University of Oxford, one of my fellow students was a lanky, bearded Australian called Geoff Cumming. Geoff was investigating the effects of 'backward' and 'lateral masking' on perception, using much the same kind of procedure as Tony Marcel. A faint image of a letter was projected on to a screen for a brief period, and then followed, after a short, variable delay, by another image, such as a checkerboard pattern. Geoff was looking at the effect of various characteristics of the second stimulus on people's ability to detect the first event: under some conditions the latter would wipe out conscious awareness of the former. During the course of his experiments, he noticed a rather curious phenomenon. When his subjects were responding at their own speed, they would, under particular combinations of conditions, regularly fail to detect the target letter. But when they were urged to respond as quickly as they could, under the same conditions, they would often make a fast response that correctly detected the target – but would then, a moment later, verbally apologise for having made a mistake! It would seem that they could break through their self-consciousness by responding very fast, and thus were enabled to make use, unconsciously, of the faint information from the first stimulus. But because this information had not been strong enough to make it through into consciousness, they retrospectively concluded that their response had been in error – and incorrectly corrected themselves.[8]

Some of the deleterious effects of self-consciousness on performance have been vividly demonstrated in certain types of brain disorder. Patients with neurological damage may show a dramatically increased disparity between their ability to function and what they are consciously aware of. Tony Marcel has recently reported the clinical case of a woman suffering from hemiplegia with anosognosia – she has lost the use of one side of her body as the result of a stroke, but seems curiously unaware of her deficit.[9] If she is asked to describe herself she will not mention the paralysis. If she is asked to rate her ability to do something that requires the use of two hands – catch a large beach-ball, for example – she will give herself eight or nine out of ten. When asked about herself directly, her consciousness of her condition is remarkably low. However, if questions are put to her in such a way that her self-image is not so directly involved, she then gives quite different answers. If, instead of asking 'How well can *you* do these things?', the questioner says 'If I were like you, how well would I be able to catch the ball?', she will give a rating of only one or two out of ten.

When the form of the question allows her to distance herself from her condition, to 'disown' it, she is able indirectly to acknowledge it. And this is not just a matter of 'not wanting to say', or feeling consciously embarrassed. The evidence is that her reluctance to admit her condition operates 'upstream' of consciousness, in the underground departments of the mind where the decisions about what to allow into consciousness are being made. Interestingly, one can create this 'disinhibition' of consciousness not only by allowing her to project her disability on to someone else, but by inviting her to talk in a regressive, childlike mode. If you hunker down by the side of her chair and whisper rather conspiratorially, 'Tell me, is the left side of your body ever *naughty*?', she will join in the game and whisper back, 'Yes, terribly.' We could speculate that there is a childish sub-personality that needs to exercise less tight control over what can be given access to consciousness – after all, children are used to not being in control of things. Much of the world for them is refractory, and 'naughty' is the childhood word *par excellence* for things that misbehave or will not do what they are told.

We know that, *in extremis*, people are capable of extraordinary feats of tactical unconsciousness. People suffering from 'hysterical blindness', for example, have cut off their consciousness of vision as a result of witnessing something horrific – to protect themselves from further trauma.[10] It is possible to raise the conscious threshold on one sensory modality, apparently, in a way that is both non-selective (unlike the 'perceptual defence' case) and extreme. Yet such people, though they have no visual experience, manage to navigate their way around obstacles uncannily well. Analogously, functionally deaf people show no response to sounds – they fail to show the normal startle reaction to a loud, sudden noise, for example – yet may respond 'No' when quietly asked if they can hear anything. Just as the patient with no memory nevertheless knows, at some level, not to shake hands with the doctor who has hurt them, so it is possible for people with no vision to be able to find their way about, and for people with no hearing to respond to questions.

We are all familiar with more mundane versions of the same phenomena, such as 'unconscious driving', which I used as an example in the last chapter. Though the idea of 'unconscious seeing' may strike us at first as weird or paradoxical, it does so only because it conflicts so strongly with our implicit beliefs about how the mind works, and this dissonance simply makes us unaware of how much of the time we respond appropriately in the absence of conscious awareness. If we were 'seeing unconsciously' in the absence of any conscious experience at all – if we were able magically to find our way around a strange house in pitch darkness, as it were – we would indeed be stunned. But our reliance on the undermind is conveniently masked by the fact that the 'hole' in consciousness is almost always plugged by some other content that is drawing our attention. It is only because our conscious mind is occupied with something else that we usually fail to observe the extent to which the visual information on which we have been relying has bypassed conscious awareness.

We are familiar too with the effect of 'self-consciousness' on behaviour as well as perception. Think of someone who is being interviewed for a job they desperately want, or a child who has been specifically enjoined to carry a full cup of tea very carefully. In such situations there is a sense of vulnerability, of a precarious balancing act, the successful execution of which depends on a degree of skill or control that we do not confidently possess. Thus there is anxiety and apprehension. And this leads to a coarsening of motor control, making us clumsy, in addition to the constriction of attention. Under pressure, we seize up, or 'go blank'. The interviewee fails to understand a perfectly straightforward question. The child concentrates so hard on not spilling the tea that her coordination goes, and she becomes graceless and gauche. It was the very day, in 1984, when my long-standing (and long-suffering) partner had finally finished our relationship that I dived – without any conscious suicidal intention – into the shallow end of a swimming pool and split my head open on the bottom. Losing keys, breaking plates and denting the car are similar symptoms of stress. It is when consciousness is most fiercely preoccupied, usually with a difficult and emotionally charged predicament, that it disregards information (like a large sign saying DEPTH ONE METRE) most egregiously, and siphons off from the unconscious the resources that it needs to function well.

One way in which the relationship between consciousness and the undermind can be radically transformed is through hypnosis (a phenomenon the existence of which, unlike the paranormal, is now established beyond empirical doubt). The active ingredients in hypnosis are relaxation and trust: allowing yourself to be hypnotised involves giving up your normal sense of controlling your own actions, of planning and striving, and putting yourself in someone else's hands. And in this state, for those who can attain it, the relationship between conscious and unconscious becomes unusually labile and permeable. The hypnotist is able to speak directly to the undermind, and is able to adjust, sometimes to an extraordinary extent, which aspects of the unconscious gain access to consciousness, and which do not. Under so-called 'hypnotic age regression', for example, you may gain conscious access to long-forgotten childhood memories. Or you may have compelling hallucinations that you take to be 'reality'. While on the other hand you may be rendered functionally deaf or blind, either to particular kinds of events, or across the board. In one case the conscious threshold is lowered so that normally inaccessible memories surface into conscious awareness; in the other the threshold is raised so high that even mundane experience is blocked out.

Though consciousness may be drastically altered or reduced, we can show that the undermind continues to function. The fact that you no longer have any conscious experience of hearing, or of pain, for example, does not mean that you are *really* not registering the sensations. The threshold between conscious and unconscious has simply been raised to the point where consciousness just isn't getting the usual reports of what is going on in the interior. Take the control of pain. Hypnotic analgesia is a reliably documented and effective method of pain control.[11] Studies show it compares favourably with drugs such as aspirin, diazepam (Valium) and morphine in reducing the experience of pain. Hypnotic suggestion alone produces clinically significant pain relief in as much as 50 per cent of a sample of the general population, even when the people in the sample have not been preselected for their hypnotic susceptibility.

Yet, despite the dramatic alteration of the conscious experience, some responses to the painful stimuli remain. The registration of pain can be demonstrated through physiological measures, for example. One widely used indicator of general arousal is the 'galvanic skin response', GSR, a measure of the resistance of the skin to the passage of electrical current. People who, as a result of hypnotic analgesia, show no visible reaction to a painful electric shock nevertheless show GSR reactions that are typical of the more normal response.[12] Also, it turns out that you can talk to a 'hidden part' of the person who can tell you about the pain, even though its conscious intensity is reduced or non-existent. If hypnotised subjects are asked to sit with their left hands in a bucket of iced water – which is normally quite painful – they will appear relaxed and will say that they genuinely feel little or no discomfort. However, if they are asked to respond with the other hand to a written list of questions about their general physical state, as it were inadvertently, they will report, in their answers, the pain that they do not 'feel'.

One of the clearest demonstrations of this so-called 'hidden observer' effect was recorded by Ernest Hilgard, a long-time hypnosis researcher, in a student practical class. One of the students, a suitable subject, was rendered functionally deaf: he denied hearing anything and failed to flinch at loud noises. While he was in this state, Hilgard whispered softly in his ear:

> As you know, there are parts of our nervous system that carry on activities that occur out of awareness, like circulating the blood . . . There may be intellectual processes also of which we are unaware, such as those that find expression in dreams. Although you are hypnotically deaf, perhaps there is some part of you that is hearing my voice and processing the information. If there is, I should like the index finger of your right hand to rise as a signal that this is the case.

The finger rose, and the hypnotised student spontaneously commented that he felt his index finger rise, *but had no idea why it had done so*. Hilgard then released the student from the hypnotic deafness, and asked him what he thought had happened. 'I remember', said the volunteer, 'your telling me that I would be deaf at the count of three, and would have my hearing restored when you placed your hand on my shoulder. Then everything was quiet for a while. It was a little boring just sitting here, so I busied myself with a statistical problem I was working on. I was still doing that when suddenly I felt my finger lift.'

Our security is threatened by information that is painful, or predictive of pain, but it is discomfited by more than that. We possess a whole variety of beliefs, many of which are themselves unconscious or unarticulated, which specify, more or less rigidly, and in more or less detail, our character and our psychology. They define what kind of person we are, our personality or 'self image', and even how our minds are supposed to work. Could it be that our consciousness, what we are able to feel and know about ourselves, is regulated by these beliefs, as well as by the need to protect self-esteem? Is there any evidence that the threshold and the nature of conscious perception – the information in the undermind which is consciously available to us – is influenced by such assumptions?

An intriguing study by Ellen Langer at Harvard suggests that even such a basic psychological attribute as our visual acuity is determined by who we happen to believe ourselves to be. Her subjects were invited to 'become' air force pilots for an afternoon. They were dressed in the appropriate uniform and given the chance to 'pilot' a jet airplane on a flight simulator. The context was made as 'real' as possible, and the subjects were asked to try to *become* a pilot, not merely to act the part. At the beginning of the study, before the simulation had been introduced and explained, each subject was given a short physical examination, which included a routine eye test. During the flight simulation, while they were being pilots, they were asked to read the markings on the wings of another plane that could be seen out of the cockpit window. These markings were actually letters from an equivalent eye chart to the one used in the physical. It was found that the vision of nearly half of the 'pilots' had improved significantly. Other groups of subjects, who were equally aroused and motivated, but who were not immersed in the role, showed no such improvement. By changing the sense of self, more precise sensory information can become available to consciousness.[13]

The self also contains core assumptions about psychological aspects of our make-up that are generic and cultural as well as those that identify us as individuals. Some of these core beliefs concern consciousness itself: when conscious awareness occurs, what it is for, how well it can be trusted and so on. And one of these tacit assumptions is that 'What we see – consciously – is what is "there", and it is *all* that is there'. If we subscribe to this model of perception, we will unquestioningly assume that 'When I have no conscious visual experience, I cannot have registered anything about what is happening out there in the visual world'. So if by some tragic accident I were to be deprived of conscious visual *experience*, but not of unconscious visual information, it is possible that I would be prevented *by this belief* from reacting appropriately to visual *events*. Someone who is involuntarily deprived of conscious sight may thus, as a result of their fundamental belief that consciousness and perception are the same thing, handicap themselves further by cutting themselves off from their residual unconscious visual capacity.

It has recently been suggested that this is exactly what may be happening in cases of so-called 'blindsight'. These cases have now become quite celebrated – they have taken over from 'split brain' patients as the most fashionable and intriguing neurological curios. The blindsight condition results from damage to the visual area of the cerebral cortex which leaves the patient with a blind 'hole' in some part of their visual field. Despite the lack of conscious visual awareness, it has been convincingly demonstrated that these patients *can* react appropriately to stimuli that impinge within the hole – but only if they are allowed to feel that they are playing a rather bizarre kind of 'guessing game', and are not actually being asked to 'see' anything. In the original studies by Lawrence Weiskrantz at Oxford, patients were asked to indicate when they saw one of a constellation of small lights that were flashed at different locations in the visual field. Those lights that fell within the blind area were, as you would expect, not reported. However, when asked to do something absurd – to take part in the nonsensical game of pointing at the hypothetical location of the (to the patient) non-existent light – they were able to do so with remarkable accuracy and consistency. Exactly *what* these patients can respond to is still under investigation. They can certainly point accurately at flashes of light, distinguish between simple shapes such as circles and crosses, and it has been claimed that two such patients have been observed to adjust the movement of their hand appropriately as they reach out for different objects which they cannot 'see'.[14]

Though we can demonstrate that blindsight patients do possess this residual visual capacity, they seem (like Tony Marcel's experimental subjects) unable to use it to respond verbally to the flashes of light, and they do not seem to make spontaneous use of this information to further their own everyday purposes. In the 1993 CIBA symposium on 'Experimental and Theoretical Approaches to Consciousness', psychologist Nicholas Humphrey, commenting on one of the presentations, had some perceptive comments to make on exactly this issue.

> John Kihlstrom's interesting remarks about the self and its relation to unconscious processes . . . are quite well supported by some of the data from patients with blindsight . . . I worked many years ago with monkeys who had had the striate cortex [the primary visual part of the brain] removed: they retained extraordinarily sophisticated visual capacity, much better than anything which has yet been discovered in human beings with lesions of the striate cortex. One way of thinking about this is that the monkey has an advantage in that it doesn't have a particularly highly developed concept of self. Hence the monkey's non-sensory [i.e. unconscious] visual percepts are nothing like so surprising to the monkey as to the human. For a human to have a percept which isn't *his own* percept (related to himself) is very odd indeed. So human patients retreat into saying: 'I don't know what's going on' and denying their ability to see at all. For the monkey, I suspect [unconscious] perceptual information doesn't create the same sort of existential paradox, therefore the monkey is much more ready to use it . . . Interestingly, for one particular monkey I worked with for a long time, there *were* conditions under which she became unable to see again – if she was frightened or she was in pain. It was as though *anything which drew attention to her self undermined her ability to use unconscious percepts.*[15] (Emphasis added)

The more self-conscious we are – the more fragile our identity – the more we shut down the undermind. As people feel increasingly vulnerable, so their access to, and reliance on, information that is faint or fleeting declines. They become not just physically but also mentally clumsy, losing access to the subtler ways of knowing. Conversely, the less self-conscious we are, the more 'at home in our skins and our minds', the more it seems we are able to open ourselves to the undermind, and to the mental modes through which it speaks. Self-consciousness is a graded phenomenon; there are milder, more chronic and more widespread degrees of self-consciousness, in which the kinds of deleterious effects we have been discussing still occur, albeit in less intense forms. Extrapolating from the experimental studies, we might hazard the suggestion that many people (at least in busy, d-mode cultures), much of the time, are in a state of low-grade, somewhat pressurised self-consciousness, and that in this state, consciousness is edited and manipulated so that its contents are as congenial and unthreatening to the operative model of self and mind as possible. We attenuate our contact with the iridescent world of the undermind, and may thus deprive ourselves of valuable data. Though we are in fact sensitive to the shimmering reality that underlies consciousness, we act as if we were not – because we do not 'believe' in it, do not trust it, or do not like what it has to say.

Blindsight research suggests that one area in which we might expect to see the effects of 'self' on consciousness is where people are acting deliberately or intentionally, as opposed to accidentally, playfully or impulsively. The sense of self is associated most crucially, after all, with plans and actions which are designed to serve our own conscious purposes. Intentions are the conscious expressions of our valued goals – of our selves, in other words. One of the features of the blindsight syndrome seems to be the decoupling of patients' unconscious seeing from their own intentions. We can show that they have residual sight, and we can do so best under precisely those conditions when they are *not* acting on the basis of any internally generated intention: when they are not trying to achieve anything, or prove themselves in any way.[16]

This inhibiting effect of intention certainly has its parallels in everyday life. The phenomena of 'not being able to see for looking', or of 'trying too hard', are commonplace. Perhaps the presence of a strong intention locks consciousness too firmly into a predetermined framework of plans and expectations, so that other information, which could potentially be useful or even necessary, is relegated to unconscious processes of perception, where it is, in these cases, ignored. *Intention* drives conscious *attention*, to the detriment, sometimes, of intelligence. In d-mode, we are not just 'looking', we are looking *for*; and what we are looking for has to be, to an extent, pre-specified. Attention is focused and channelled by the unconscious decisions we have made about what may be 'relevant' to the solution of the problem, or the achieving of the intention. And these presumptions may be accurate, or they may not.

Sigmund Freud made exactly this point in his 'Recommendations to physicians practising psychoanalysis', published in 1912. The technique of psychoanalysis, he said,

> consists simply in not directing one's notice to anything in particular, and in maintaining the same 'evenly-suspended attention' . . . in the face of all that one hears. In this way . . . we avoid a danger. For as soon as anyone deliberately

> concentrates his attention to a certain degree, he begins to select from the material before him; one point will be fixed in his mind with particular clearness and some other will be correspondingly disregarded, and in making this selection he will be following his expectations or inclinations. This however is precisely what must not be done.[17]

This line of thought suggests that threat or desire may make consciousness narrower as well as coarser, and may explain the experiments described in Chapter 5 which showed that the creativity of intuition and problem-solving is reduced by a feeling of threat or pressure. One of the major reasons why too much effort, too purposive an attitude, or a general increase in stress or anxiety is counterproductive is because it creates 'tunnel vision'. We might imagine that, at any given moment, people are shining a 'beam' of attention outward on to their environment, through the five senses, and inward, too, on to their own physiological, emotional and cognitive state. At the extremes, this beam may be tightly focused, like a spotlight, or wide and broad, like a floodlight, or the dim glow of a candle.

In principle, both forms of attention, concentrated and diffused (and all the degrees of focus in between), are useful. In a pitch-black cave, a hurricane lamp shedding a broad, dim light, which enables you to see the overall size and shape of your surroundings, is what you need first. If all you have is a torch with a fine beam, you will not be able to get your bearings so well. But once you have orientated yourself, it is useful to be able to home in on details, and now the spotlight comes into its own. The diffuse illumination gives you a holistic impression; the focused beam enables you to dissect and analyse. Both are needed, and in an optimal mental state one can flow between the two extremes, adopting a degree of concentration that is appropriate for that particular moment. This balance between perception and attention has a parallel in the equilibrium of deliberation and intuition required for optimising our creative thought processes.

It appears that being stressed, threatened or over-eager tends to narrow the beam of attention too much, whether it be to one's own internal database or to the outside world. People who are chronically anxious have been shown to have more tightly focused attention than people who are more relaxed. Several investigators have reported impaired night vision, for example, in people who are stressed or anxious.[18] When people are required simultaneously to carry out a focal task, such as trying to track a randomly moving point in the central part of a screen with a cursor, and a task that requires the use of peripheral vision, like detecting small flashes of light at the edges of the screen, then increasing the size of the reward for successful performance leads to a tighter concentration on the tracking task, and a serious fall-off in performance on the peripheral task. And if subjects had *not* been forewarned about the peripheral lights, 34 per cent of subjects who were working for large rewards failed to notice them at all, whereas only 8 per cent of those receiving small incentives failed to notice them.[19] The same kind of tunnel vision is produced if the general level of stress is increased by making people work in hot or noisy conditions.[20]

People working under pressure, whether environmental or psychological, tend to select out and focus on those aspects of the situation as a whole which they judge to be the crucial ones. And this judgement must to a certain extent, as Freud realised, be a prejudgement. You make an intuitive decision about what is going to be worth paying attention to. If this 'attentional gamble' is correct, people may learn the task, or figure out a solution, quicker, but at the expense of a broader overview. They see in terms of what they expect to see, and if this self-imposed blinkering reflects an adequate conceptualisation of the problem, time may be saved. But if it is not, or if (as in the Luchins' jars experiments) the situation changes but because of the tight focus the change is not noticed, then a commitment to the spotlight processing strategy is going to let them down. As Jerome Bruner says, reflecting on the adverse effect of motivation: 'Increase in incentive leads to a higher degree of selective attention for those parts of a complex task that subjects interpret as more important, with a concomitant tendency to pay less attention to other features of the situation.'[21] Broad, diffuse attention is precisely what is needed in non-routine, ill-defined or impoverished situations, where data is patchy, conventional solutions don't work, and incidental details may make all the difference. And that is why too much effort inhibits creativity.

An experiment by Jerome Singer demonstrates how increasing the desire for a solution can lead to the coarsening of perception. He asked subjects to estimate the size of a square placed at some distance away from them down a long corridor, by selecting one from an array of different-sized squares arranged on a stand to one side of them. Although this looks simple enough, it is in fact quite a difficult task, precisely because there is so little information with which to judge the size of the distant square. All kinds of subtle cues, such as shadow, brightness and the visual texture of the square, might be helpful. So though the square occupies a very precise point in the visual field, subjects will benefit from having a wide beam of attention in terms of the range and kinds of cues to which they are attentive. It is the sort of task, in other words, which might prove sensitive to the effects of pressure. When subjects were required not just to make their judgement but to imagine that they had a bet riding on its accuracy, their performance deteriorated – even with an imaginary as opposed to a real stake. In another version of the study, subjects were asked to spend fifteen minutes, prior to the size test, working on an insoluble problem, with the experimenter feigning surprise and disappointment at their poor performance. This was sufficient to induce a mood of anxiety and frustration which, in its turn, coarsened the perception of the cues in the distance test, and caused a deterioration of performance.

CHAPTER 9

The Brains behind the Operation

The Brain – is wider than the Sky –
For – put them side by side –
The one the other will contain
With ease – and You – beside –

Emily Dickinson[1]

We now know a considerable amount about what the intelligent unconscious can do, and the conditions under which it works best; but it remains to discover what it is – how it is physically embodied – and exactly how it works: how it makes available the slower ways of knowing and the powers of subliminal perception. We know very clearly that there is a 'brains' behind the operation, but who, or what, is it?

The brain – the three pounds of soft wrinkled tissue that occupies the skull – is the focus of intense research activity at the moment. The 1990s were designated the 'Decade of the Brain' by the US Congress. At the 1996 British Association for the Advancement of Science 'Festival of Science', the annual public showcase for the work of scientists in Britain and elsewhere, the two-day symposium on 'Brains, Minds and Consciousness' had to be moved from its original venue to the largest lecture theatre on the University of Birmingham campus in order to accommodate the audience. Scarcely a month goes by without the appearance of another book by one of the leading figures in the thriving new discipline of 'cognitive neuroscience'. In trying to understand the physical substrate of the slower ways of knowing, the brain is clearly the most fruitful place to start. If we first explore what the brain does, or what it can plausibly be supposed to do, we shall then be able to see what, if anything, is 'left over', in need of explanation by some other means.

The brain is one of three main systems that coordinate the workings of a whole animal. Together with the hormonal system and the immune system, the central nervous system, of which the brain is the headquarters, ensures that all the different limbs, senses and organs of the body act in concert.[2] The brain integrates information from the eyes, ears, nose, tongue and skin with data about the state of the inner, physiological world, and, by referring this information to the stored records of past experience, is able to construct actions that respond as effectively as possible to the current situation. The brain assigns significance, determines priorities and settles competing claims on resources, for the common good. It ties together *needs*, as signalled from the interior, *opportunities* (and threats) in the environment, as flagged by the five senses, and *capabilities*, as represented by the programmes that control movement and response. And it is able to do this, in the case of human beings, with such consummate elegance and success because it remembers and learns from what has happened before.

The brain is composed of two types of cells – glial cells and neurons – both in profusion. The glia seem to be mainly responsible for housekeeping: they mop up unwanted chemical waste, and make sure that the brain as a whole stays in optimal condition. But it is the neurons, approximately one hundred billion of them, that give the brain its immense processing power. Each neuron is like a minute tree with roots, branches – the dendrites – and a trunk called the axon. The neurons in the brain vary considerably in their actual size and shape. Some are straggly and leggy, with long axons running for some millimetres through the brain; others are short and bushy, with dense dendritic branches, but perhaps measuring only a few thousandths of a millimetre from end to end. However, all the neurons have the same function: they carry small bursts of electric current from one end to the other.

The neurons are packed together into a dense jungle, and where their roots and branches touch (at junctions called synapses) they are able to stimulate one another. Electrical activity in one – what we might call the 'upstream' neuron – influences the likelihood that its 'downstream' neighbour will become electrically active in its turn. Normally each downstream neuron needs to collect the stimulation from a number of its upstream neighbours, until it has built up enough excitation of its own to exceed some 'firing threshold'. When it fires, a train of impulses is initiated along its own axon and into its dendrites, where it can contribute to the excitation of other cells with which it is in contact. No one input will fire a downstream cell on its own, but each contributes to this general pool of activation, making the cell more or less disposed to fire in response to other inputs.

The elaboration of the story of how the neural electrical impulses originate, are conveyed along the axon, and serve to activate other neurons is one of most notable successes of twentieth-century science, and it has been told in detail on many occasions. In brief, each neuron is covered with a semi-permeable membrane that is able to retain some kinds of chemicals within the body of the cell and keep others out. Many of these chemical particles carry a small electrical charge, either positive or negative, and the membrane, in its normal state, is selectively permeable to these 'ions', in such a way that it is able to maintain an electrical gradient, a potential difference, between the inside of the cell and the fluid which surrounds it. However, under the influence of other chemicals – neurotransmitters – which may be released into the ambient fluid, the character of the membrane changes so that charged ions are allowed to flow across it, and it is this flow which initiates the chain of events that may result in a burst of electricity, an action potential, travelling from one end of the neuron to the other.

Action potentials occur spontaneously at more or less regular intervals: nerve cells are never completely quiet. Even when we are asleep their activity continues. But the pattern and frequency of firing can be dramatically altered by events at the synaptic junctions with other cells. A wave of electricity arriving at a synapse from an upstream cell causes neurotransmitters to be released into the gap between it and the downstream cell. These molecules float across the gap and attach themselves to receptors on the membrane on the downstream side, causing it, in its turn, to allow charged ions to flood into the cell and set off another action potential. The stimulation that one neuron gives to another can be inhibitory, making the next-door cell less likely to fire, as well as excitatory.

Each cell may receive stimulation from up to 20,000 different sources, so the neuronal jungle as a whole is incredibly tightly and intricately interconnected. It is estimated that there are of the order of one million billion possible connections in the outer mantle, the cortex, of the brain alone. If we could lay the brain out neatly, we would see that on one 'side' there are all the incoming 'calls' from the inner and outer senses; on the other there are the outputs and commands to all the muscles and glands of the body; and in the middle is this vast tangle of living wires, immersed in a complex, continually changing bath of chemicals, integrating and channelling

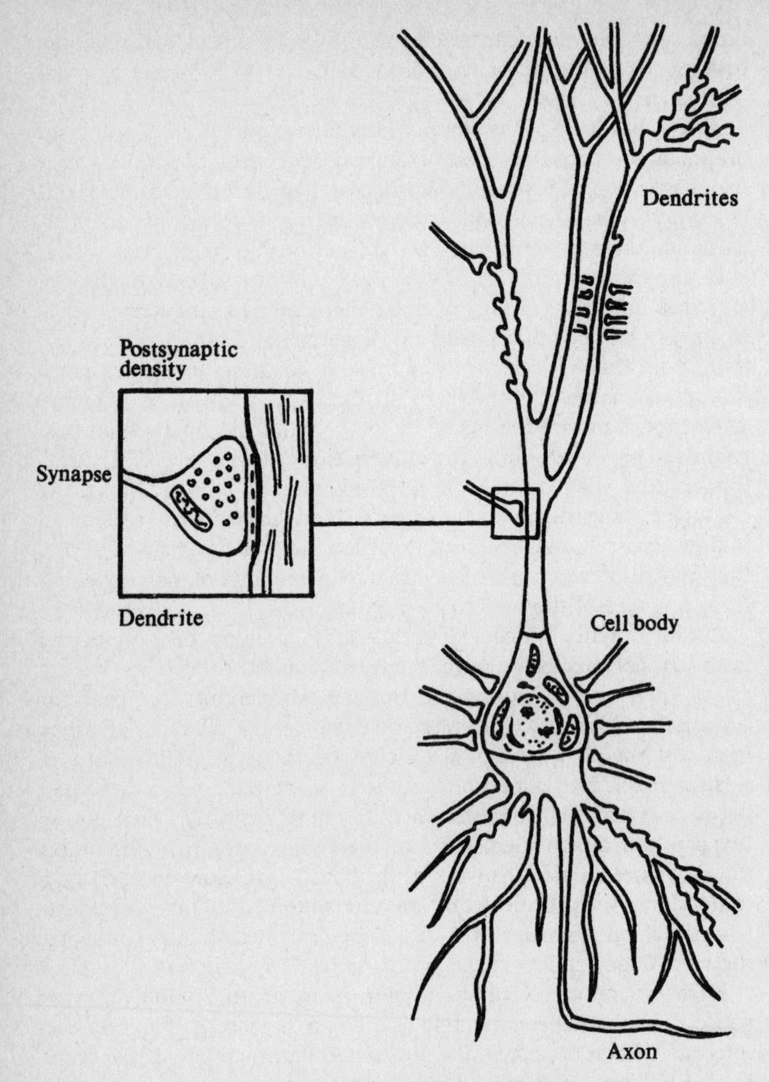

Figure 8. A stylised neuron
(reproduced with permission from Scientific American*)*

messages from one side of the network to the other, with many, many loops and diversions on the way.

The electrical communication between neurons can be changed permanently as a result of experience. Cells that were originally strangers, and relatively 'deaf' to each other, can develop close associations, as a result of which one only has to whisper to attract the other's attention where before it had to shout. One way in which experience affects the long-term flow of neural communication is through physical dendritic growth. It has been shown that animals who live in environments full of rich stimulation develop much bushier neurons than do those whose worlds are dull and monotonous. The total number of synapses can increase. But synapses are also capable of becoming easier for an impulse to cross – and for this we need a different process: 'long-term potentiation' (LTP).

When neurotransmitters are released into the synaptic channel between the upstream (currently active) and downstream (currently inactive) neurons, some of the pores in the downstream membrane open up easily to let the charged ions across. However, others, the so-called NMDA (N-methyl-d-aspartate) receptor sites, start out by being more tightly constricted, and will only open up if they are subjected to stimulation that is strong and long-lasting. But once they *have* been opened, they take less persuading on subsequent occasions. For some time after they have been subjected to strong stimulation, the NMDA pores will respond to a much weaker signal. This is one of the fundamental mechanisms that allows the brain to learn.[3] As one of the pioneers of brain research, Donald Hebb, wrote in his seminal book *The Organization of Behavior* in 1949, 'When an axon of cell A is near enough to excite a cell B and repeatedly and persistently takes part in firing it, some growth process or metabolic change takes place in one or both cells such that A's efficiency, as one of the cells firing B, is increased.'[4] Or, more informally, 'cells that "glow" together, grow together'.

An important characteristic of LTP is its *specificity*. Though one neuron receives inputs from a myriad of other cells, if, through LTP, it becomes more responsive to one particular upstream neighbour, it does not become indiscriminately more intimate with the others. Thus there are mechanisms in the brain which enable specific paths of facilitation to be developed between groups of neurons. When a human baby is born, there are certain genetically determined frameworks that serve to impose a general structure on the brain, but much remains unfixed. The brain is like a vast lecture theatre of students on their first day at university: full of potential friendship, but as yet strangers to each other. After a few weeks, however, each student has begun to belong to a number of different evolving circles of acquaintance: study groups, sports teams, neighbourhoods and clubs. Just so, every neuron in the brain comes to belong to a variety of developing clusters, each of which is bound together in such a way that stimulation of one member, or a small group of members, is likely to lead to the 'recruitment' of the others. And the major reason why they tend to become associated is simply that they are active at the same time. Communication becomes more selective, and information begins to flow, throughout the neural community as a whole, along more stable channels.

In order to begin to make links between our understanding of mind and brain, we need to think in terms of the behaviour of these clusters or assemblies of neurons, not of the individual nerve cells themselves. Though we need the molecular level to help us understand how assemblies of cells are formed, and how they behave, there are properties of large assemblies of neurons that are not reducible to their biochemistry – just as the way a hockey team functions cannot be understood from even the most detailed observation of a single player. Nor could the behaviour of an individual student in the context of her hockey practice be predicted or explained on the basis of her performance in the chemistry laboratory, or the college bar. Though there is some direct evidence about how assemblies of neurons behave, it is much more difficult to gather, and still relatively sparse compared with what we know about individual cells. We have to reach out beyond established fact into the realm of plausible hypothesis; to use what we know as a basis on which to build more holistic images of what the brain is like and how it works.

Suppose you are shown dozens of photographs of the same person, Jane, in different moods, wearing different clothes, in different company, engaged in different activities. Some of Jane's features will stay the same across all the photos: the colour of her eyes, the shape of her nose, and more hard-to-pin-down constellations of the way her features are put together. You may not be able to say what it is about the pattern of Jane's face, but after a while you would recognise her anywhere. The neural clusters that correspond to these 'core features', those that recur every time you see Jane, will always be co-active, and it is they that will therefore bond most tightly together. They become the nub of your concept of Jane. Others of her features, such as her ready smile or her penchant for floppy hats, while not critical, become closely associated with her representation, so that, in the absence of any definite information to the contrary, you may automatically fill in these default characteristics when you think of Jane.[5]

More loosely associated still, there are a whole variety of features and associations that have been connected with Jane, but which are less characteristic or diagnostic of her: the time she was photographed with chocolate ice cream all round her mouth; the scarlet suit she wore to Tim and Felicity's wedding. These memories, we might say, form a neural penumbra that is activated a little when we are reminded of Jane, but not – unless the context demands it – strongly enough for them to fire in their own right. Thus the overall neural representation of Jane is not clear-cut. It is composed of an ill-defined, fuzzy collection of features and associations, some of which are bound very tightly together at the functional centre of her neural assembly, while others are more loosely affiliated, and may on any particular occasion form part of the activated neural image of Jane, or may not. Some of these features, whether core, default or incidental, may be distinguishable, or nameable, in their own right – 'nose', 'smile' – while others may comprise patterns that are not so easy to dissect out and articulate.

Depicting the concept of 'Jane' as a vast collection of more or less tightly interwoven neurons does not commit us to supposing that there is a single 'Jane' neuron anywhere in the brain to which all the others lead, or even that the 'extended family' of neurons must all be found in the same location within the brain. There is plenty of evidence that such families of neurons can be – indeed usually are – widely distributed across the brain as a whole. Even if we focus only on the world of sight (and concepts are generally multi-sensory) we find that different aspects of vision – colour, motion, size, spatial location – are processed in quite widely separated areas. Current estimates suggest that there are at least thirty to forty discrete areas of visual processing in the brain, and that these different systems of neurons are themselves interconnected in highly intricate ways. Add in the other senses, as well as memory, planning and emotion, and there will be traces of 'Jane' in every corner of the brain. Just as, in the modern world, geographical proximity is only loosely indicative of the strength of people's relationships, so the intimacy between neurons is reflected in their functional, and not necessarily their physical, closeness.

Since before birth, experience has been constantly binding the brain's neurons together into functional groupings which act so as to attract and 'capture' the flow of neural activation. And these centres of activity in their turn become strung together to form pathways along which neural activation will preferentially travel, so that the brain as a whole develops a kind of functional topography. To explore the consequences of this idea, we might imagine that a concept like 'Jane' or 'cat' or 'student' forms an activation 'depression' towards which neural activity in the vicinity will tend to be drawn, as water finds its way into a hollow. Experience wears away bowls and troughs in the brain which come to form 'paths of least resistance', into and along which neural activity will tend to flow.

At the bottom of a hollow are the attributes that are most characteristic of its concept: those by which, whether we know it or not, the concept is recognised. On the sides of the valley lie the default properties, and higher up are those associations that are optional or incidental. Experience erodes and moulds the mass of neurons into a three-dimensional 'brainscape' where the 'vertical' dimension indicates the degree of functional interconnectedness, the mutual sensitivity and responsivity, of the neurons in that conceptual locality. The deeper the dip, the more tightly bound together the neurons; the more 'deeply engrained', we might say, that concept, that way of segmenting reality, is. But hollows vary in their steepness of incline as well as their depth and their size. Valleys that have steep sides are those where the concept being mapped is well defined: it has relatively few associations that are not criterial. Gentle slopes indicate a wider range of looser connections and connotations.

It is technically impossible to examine in a living brain how such large-scale, distributed collections of neurons come to be associated in the course of everyday kinds of learning. However, it is possible to write computer programs that simulate the properties of neurons, and explore the learning that relatively small numbers of such artificial neurons can accomplish. It turns out that these so-called 'neural networks' are remarkably intelligent. They can, for instance, mimic very closely the kinds of learning that we discussed in Chapters 2 and 3, where complex sensory patterns are picked up and transformed into expertise, without any conscious comprehension or explanation of what has been learnt.

Take as an example the problem of using echo-sounding equipment such as Asdic and sonar to detect mines at sea. The need to discriminate between underwater rocks and sunken mines is obviously a pressing and practical one, both during naval warfare and in the clean-up operations that follow it. Yet it is not an easy problem to solve, for a number of reasons. Echoes from the two types of object can be indistinguishable to the casual ear. And the variations *within* each class are massive – both rocks and mines come in a wide variety of sizes, shapes, materials and orientations – much greater, apparently, than the differences *between* the two. If there are any consistent distinctions, they are almost certainly not to be made in terms of single features, such as the strength of a signal at a specific frequency, but will involve a variety of patterns and combinations of such features.

Suppose we were to analyse any particular echo into thirteen frequency bands, and to measure the amplitude of the signal in each of these bands. Call the bands A, B and so on up to M; and say, for the sake of argument, that the signal strength in each band could range from 0 to 10. It is unlikely that any one of these bands on its own would provide us with a decisive fingerprint. The solution to the problem of discrimination is not as simple as saying that all rocks score more than 7 in band H and all mines score less than 7. It is not even as simple as saying 'If the echo came from a rock, then its strength in band C will be between two and three times its strength in band J'. The only kind of pattern that might conceivably distinguish rocks from mines will be something like: 'An echo probably comes from a mine if *either* the total value on bands A, D and L exceeds the product of the value on bands E and F by more than a factor of six, at the same time as H minus K is less than half of J divided by B; *or* the total of G, H, K and L is more than 3.5 times the total of A, B and C divided by the difference between I and M.' To make the discrimination successfully, if it can be done at all, will require the detection of patterns of this degree of complexity: ones that are hard to describe, let alone to discover.

In fact, human operators can become quite accurate at making these judgements, though, like the subjects in the 'learning by osmosis' experiments, they cannot articulate what it is they know. (You may remember the finely tuned 'intuition' of the sonar operator in the film *The Hunt for Red October*.) Nevertheless, human beings are less than perfect at it, and mistakes are potentially costly. To learn to discriminate between rocks and mines poses an interesting real-life challenge for a simulated brain.

A neural network comprising only twenty-two different 'neurons' has learnt to perform this discrimination surprisingly well. The neurons are arranged together in three 'layers' (as shown in Figure 9). The first layer of thirteen 'sensory' neurons corresponds to the thirteen frequency bands into which the sound spectrum of the echo is divided. They are tuned to detect the signal strength

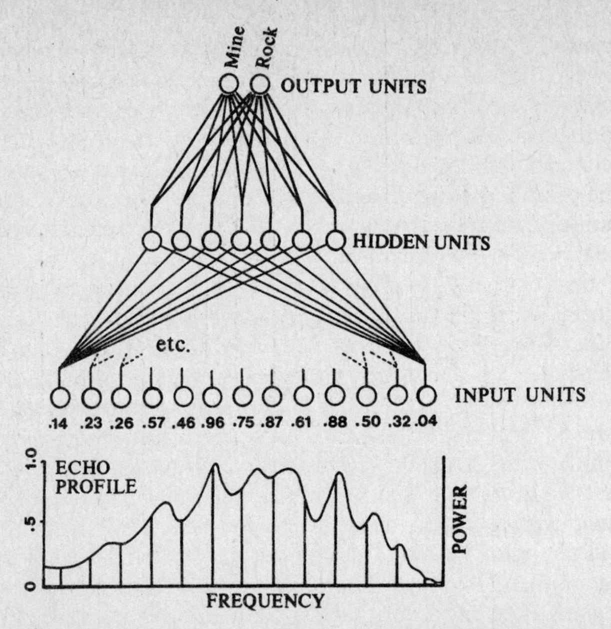

Figure 9. A simple neural net for distinguishing rocks from mines

within their particular band, and to emit a signal, like a real neuron's burst of action potentials, that is proportional to this strength. All of these sensory neurons send their signals to each of the seven neurons that are arranged in the next layer. And each of these seven starts out by sending a copy of its output to each of two units in the final layer, the output of one of which corresponds to the decision 'It is a rock' and the output of the other signals that 'It is a mine'. This simplified brain is not able to grow any more connections, but it is able to adjust the selective sensitivity of every neuron to each of the inputs that it receives, in exactly the way that real nerve cells do.

The 'job' of the network is gradually to adjust these sensitivities, in the light of experience, so that the flow of activity through the connections reliably activates the 'rock' neuron whenever it is given a rock echo, and the 'mine' neuron when it is given a mine. Neither the programmer, nor certainly the computer, knows at the outset what the requisite sensitivities are, nor even whether a set of sensitivities that will solve the problem actually exists. The best the programmer can do is to get a large and varied set of genuine sonar echoes which she *knows* arose from either a rock or a mine, and to feed these, one by one, into the network, telling it, after it has generated a decision, whether it was correct or not. In this 'training phase' the computer is given some relatively simple 'learning rules' which tell it how in general to adjust the sensitivities of the neurons as a function of its success or failure. For example, the network may be programmed to adjust all the sensitivities after each trial on the basis of their history of being associated with a correct response. Those units that have a better 'track record' are adjusted very little; those that have a poorer track record are adjusted by a larger amount. Finally, after the 'brain' has been given a large number of such feedback sessions, where it is 'told' whether it was right or wrong, it can be tested with a new set of echoes it has never met before to see what judgement it now makes.

In this example the network behaves exactly like the human subjects in the 'learning by osmosis' experiments. Simple neural networks turn out to be excellent models for this kind of learning. The network starts out 'guessing' and making many mistakes; but gradually its performance improves until finally it is capable of distinguishing quite accurately between rock and mine echoes which it has never heard before. These simulations show convincingly that brains can do what people do; that is, detect intricate, unverbalised patterns that are embedded within a wide range of seemingly diverse experiences, and use these to guide skilful action. Neither the real-life human being nor the artificial brain 'knows' what it is doing, nor on what basis it is doing it. Their 'knowledge' – successful, sophisticated knowledge – is contained in small adjustments in the way the neurons of the brain respond to each other; adjustments which simply direct the flow of activation along different channels, and combine it in different ways. All the brain needs is a diet of training experiences, some feedback, and clear, unpremeditated, unpressurised *attention* to what is happening; its intrinsic operating characteristics will do the rest.

It is worth noting that, in the rocks and mines simulation, the artificial brain came to make the discrimination with a degree of accuracy that even surpassed that of experienced human sonar operators on a long tour of submarine mine-sweeping duty. The neural network, despite its simplicity, outclasses a human expert – not because the computer is 'cleverer', but just because we have not been equipped by evolution with ears sensitive enough to divide sonar echoes into so many frequency bands. We might confidently suppose that if the same kind of problem were to use, instead of a range of metallic 'pings', human babies' cries denoting either 'hunger' or 'wind', mothers would outperform the computer comfortably. Conversely we might expect to find that a dolphin could be trained to beat both the computer and the human operator on the rocks and mines problem.

The imperfect performance of human beings reminds us that there are, of course, limits to the complexity that the unencumbered brain can handle. The world must contain many subtle contingencies that even the fine tuning of the human brain cannot pick up – especially those that have not in the past been directly relevant to survival, or which embody new technological, pharmaceutical or sociological patterns which the biological receiving apparatus was not designed to detect. And also there are many situations which we might *like* to master where there simply is no useful information, no pattern, to be picked up. But what is clear is that *the fundamental design specification of the unconscious neural biocomputer enables it to find, record and use information that is of a degree of subtlety greater than we can talk or think about.* If we let our view of the mind as predominantly conscious and deliberate blind us to the value, or even the existence, of unconscious ways of knowing, we are the poorer, the stupider, for it.

The brain works by routing activity from neural cluster to neural cluster according to the pattern of channels and sensitivities that exist at any moment, and that is all it does. Just as a pebble thrown into a pond starts an outward movement of concentric ripples, so activity in one area of the brain forms what Oxford neuroscientist Susan Greenfield calls an 'epicentre' from which activity spreads out, interacting with other flows of activation, and triggering new epicentres, as it goes. One can literally watch it happening. Studies by Frostig, Grinvald and their colleagues in Israel have used special dyes that can be introduced into cortical neurons, and which fluoresce when the cell becomes electrically active. If a spot of light is flashed into an animal's eye, a neuronal cluster can be seen to form instantaneously and may double its size in a matter of 10 milliseconds. After 300 milliseconds there may be a very large group of active cells distributed over a wide area.

The distributed nature of the neural clusters has been demonstrated by Wolf Singer in Germany. Singer has found that neurons that are widely separated across the visual cortex can nevertheless synchronise their firing patterns in response to a stimulus. Thus, as I have already suggested, the flow of activity is not literally from place to place, but between distributed patterns that continuously segue into one another. The brainscape is, as I have argued, delineated functionally, not physically. If we were able to track the brain's activity, and simplify it, we should see something that looked not like a brightly lit train travelling at night, but more like a luminous kaleidoscope being continually shaken. But to show these iridescent patterns shimmering across the brain is beyond our technical capacity, not least because they move so fast. Ad Aersten and George Gerstein have shown that neuronal groups are highly dynamic, forming and reforming within periods as short as a few dozen milliseconds. And what is more, the same neuron may take part in different patterns from moment to moment. Despite the huge technological problems, there is already some direct evidence for the existence and the properties of these neural patterns.[6]

As well as long-term, 'structural' changes in the brain, there is a variety of shorter-term influences on its responsiveness as well. The topographical 'erosion' of the brainscape is heavily modulated by much more transient influences. Brain responses are affected by the state of need, for example. If an animal is hungry, thirsty, sexually aroused or under threat, groups of neurons tend to adopt the same pattern of firing – to work in synchrony, in other words – more than when the animal is relaxed and sated. Heightened arousal seems to encourage groups of neurons to bind more tightly together into functional teams, and this, Susan Greenfield argues, has a number of interesting consequences, in addition to making each such group more excitable.

Given that neurons are linked together by both excitatory and inhibitory connections, increased arousal can have a mixed effect, causing some neighbours of an active cell to fire more readily while effectively suppressing others. In particular, there tends to be what is called *reciprocal inhibition* between a group of neurons that is currently active, and others that lie outside the group, and the extent of this inhibition makes for a more or less competitive relationship between different centres of activity. When one cluster is creating a strong inhibitory surround, it will tend to suppress other potential epicentres, and at the same time it also tends to sharpen the borders of its own pattern. Instead of priming a broad range of its associates, to varying extents, inhibition makes for a clearer cut-off, and the neural repercussions of any centre of activation thus become more limited. In the presence of a drug – bicuculline – that is known to block mutual inhibition between neurons, a pool of activity can be enlarged as much as tenfold. Thus when arousal is lower, several different centres may be activated simultaneously, because the competition is less fierce; and at the same time patterns of activation that started from different centres can flow into each other like watercolours on wet paper. And from these effects, Greenfield argues, may arise a third consequence of arousal: precisely because of the greater competition, any current 'winner' is more unstable, more likely to be toppled by the next emerging epicentre at any moment, and thus the train of thought may gather greater speed.

We know some of the chemical mechanisms that underlie this 'neuromodulation' effect. The brain stem, the oldest part of the brain, forms a bulge at the top of the spinal cord, and from it bundles of neurons project up into the midbrain, and thence into the cortex. It is these neurons that underlie the role of need, mood and arousal in varying the way cortical neurons behave. They can release chemicals called amines into the cellular milieu which make synapses transiently more or less sensitive. These amines include serotonin, acetylcholine, dopamine, norepinephrine and histamine. Acetylcholine, for example, inhibits one of the normal 'braking' mechanisms that causes a neuron to turn itself off after it has been active for a while. In general, neurons and neural clusters can be primed or sensitised by the influence of amines, so that they are more on a 'hair trigger'.

The dynamics of the brain can therefore vary in a number of different ways. The *direction* of activity flow is influenced both by the sensitivity of the long-term connections between cells, but also by the extent to which different areas are primed. A weak pathway may be temporarily boosted to the point where it is preferred over one that is normally stronger – and thus the 'points' can be switched so that the train of activation is diverted on to a less familiar branch-line. The *breadth* or degree of focus of activation of a concept may be reduced or expanded, so that a familiar conceptual hollow can be made to function as if it were either more or less clearly delineated – as more stereotyped, or more flexible, than its underlying set of structural interconnections would suggest. In one mood, a pattern may have boundaries that are clear and sharp; in another, its influence may spread more widely and taper off more gradually. The number and *variety* of different epicentres that can be simultaneously active also vary. In a state of high arousal, a single chain of associations that is more conscious and more conventional will tend to be followed. In a state of relaxation, activity may ripple out simultaneously from a range of different centres, combining in less predictable ways. And finally the *rate* of flow can vary. In a state of low arousal, a weak pool of activity may remain in one area of the network for some time before it moves on or is superseded.

Under greater arousal – when threatened or highly motivated – activity may flow more rapidly from concept to concept, idea to idea.

CHAPTER 10

The Point of Consciousness

In the beginning, man was not yet aware of anything but transitory sensations, presumably not even of himself. His unconscious brain-mind did all the work. Everything man did was without understanding.

Lancelot Law Whyte

Interesting intuitions occur as a result of thinking that is low-focus, capable of making associations between ideas that may be structurally remote from each other in the brainscape. Creativity develops out of a chance observation or a seed of an idea that is given time to germinate. The ability of the brain to allow activation to spread slowly outwards from one centre of activity, meeting and mingling with others, at intensities that may produce only a dim, diffuse quality of consciousness, seems to be exactly what is required.

There is direct evidence that creativity is associated with a state of low-focus neural activity. Colin Martindale at the University of Maine has monitored cortical arousal with an encephalograph, or EEG, in which electrodes attached to the scalp register the overall level and type of activity in the brain. When people are more aroused, these 'brainwaves' are of a higher frequency, and are more random, more 'desynchronised'. When they are relaxed (but still awake), their brainwaves are slower and more synchronised: the so-called 'alpha' and 'theta' waves. Martindale recorded the EEGs of people taking either an intelligence test – one that required analytical thought – or a creativity test – one which asked people to discover a remote associate that linked apparently disparate items, or to generate a wide range of unusual responses to a question such as 'What could you use an old newspaper for?' Using a standardised questionnaire, the subjects had first been divided into those who were generally creative and those who were not. Cortical arousal was seen to increase equally for both groups when they were taking the intelligence test, relative to a relaxed baseline. When subjects were working on the creativity test, the EEG of the uncreative subjects was the same as for the intelligence test; but the arousal level of the creative people was *lower* even than their baseline control readings.

In a follow-up study, Martindale divided the creative task into two phases: one in which people were required to think up a fantasy story, and a second in which they wrote it down. He argued that the first stage, which he called the phase of 'inspiration', would rely on creative intuition, while the second, the 'elaboration' phase, would involve a more conscious, focused attempt to work out the implications of the storyline and arrange them into some coherent sequence. As predicted, those subjects who were judged to be less creative showed the same high level of arousal in both phases, while the creative subjects showed low arousal during inspiration, and high during the elaboration. In Chapter 6 I argued that the productive use of intuition required a variable focus of attention; the ability to move between the concentrated, articulated processes of d-mode and a broader, dimmer, less controlled form of awareness. Martindale's results show that this fluidity is mirrored in the physiological functioning of the brain. Creative people are those who are able to relax and 'let the brain take the strain'.[1]

The classic description of creativity divides it into four phases: preparation, incubation, illumination and verification. (Martindale's 'inspiration' corresponds to incubation and illumination, while his 'elaboration' corresponds to what in a more scientific context would be called verification.) During the preparation phase, information is gathered and analysed through focused attention or d-mode, in which the brain acts as if the neuronal clusters were relatively sharply demarcated, and trains of associations unfold in a relatively conscious, relatively conventional manner. If the problem in hand is routine this mode will suffice to generate a solution. However, if the problem is more unusual, this way of knowing will result in a series of blind alleys. Activity rushing through tightly delineated channels will not be able to spread out broadly or slowly enough to make simultaneously active the remote associations on which the creative solution rests.

But if someone is able to move into the incubation phase, the sharp inhibitory surrounds which d-mode employs to keep activity focused and corralled, and which tend to turn gentle valleys into functional canyons, are relaxed, and the wider distribution of activity across the brainscape allows a greater number of different foci to become active at the same time. The pattern of activation in the low-focus brain resembles more the one produced by a handful of gravel flung scattershot across the surface of a still lake, than the linear sequence of epicentres created by the 'skimming' of a smooth flat stone. Now if the residual activation from the earlier, preparatory stage remains – if the problem has been put to the back of the mind without being forgotten entirely – the neural clusters that correspond to the problem specification will still be primed. (Uncreative people, as well as having lost the knack of entering the low-focus state, may also be unable to retain this background level of priming: they do not know how to put something 'on the back burner' without it falling down behind the cooker.)

As creative people go about their business, the normal exigencies and incidents of daily life will keep activating thousands of concepts and clusters throughout the brain. If one of these should inadvertently facilitate a link between previously unconnected, but primed, parts of the network, there may be just enough added activation to make an image or a metaphor exceed its threshold and shoot into consciousness – producing an 'insight', an illumination. Finally, during the phase of elaboration or verification, the focus of activation may narrow again in order to explore in more detail the implications that have been opened up.

If creativity is associated with forms of brain activity that are dim and diffuse, and if this is because, in such a state, a greater number of different foci can be simultaneously active, then we might expect that forms of unconscious awareness – subliminal perception – would show the same increase in diversity of associations. Specifically, if an idea is activated unconsciously, its associative ripples may extend out more widely than if its activation is concentrated to a degree that produces clear consciousness. There is some evidence that this is the case. Recall the study by Bowers and his colleagues (in Chapter 6) in which subjects tried to find a single word that was a remote associate of each of fifteen words presented in a cumulative list. Spence and Holland have used the same type of materials to examine the effect of unconscious perception. Their subjects were given a list of twenty words to memorise, of which ten were remotely associated with a single word such as, in the Bowers example, 'fruit'. The other ten words were of the same general type and familiarity, but were not linked together in this minimal way.

Prior to learning the list, some of the subjects were exposed to the word 'fruit' presented subliminally; some saw 'fruit' presented consciously; and some were shown only a blank screen. The results showed that the subjects who perceived 'fruit' unconsciously recalled more of the associated words on the list than did either of the other two groups. Spence and Holland interpret this result to mean that having an object clearly in conscious awareness reduces the range of associates that are active in memory to its 'immediate family'; while a stimulus that does not quite reach the focus or intensity required for consciousness subliminally primes a wider circle of associations. Focal consciousness, we might surmise, is associated with the concentration of a limited pool of 'activity' within a smaller area of the memory network.

I showed in Chapter 4 that incubation also supports better thinking through allowing time for false starts and erroneous conceptualisations to fade away, and to be superseded by a different approach. We can now see how the brain makes this process of 'reappraisal' possible. Imagine that activity in the neural network is flowing along a channel and comes to a point of choice – a T-junction. Which way is it to go? Under normal circumstances, we can assume (for purposes of illustration) that all the activity has to follow the most well-established route. If one arm of the T is worn deeper, and/or is more highly primed, than the other, then that will be the path that is preferred. We can represent the relative facilitation of each pathway, as in Figure 10, by the thickness of its line. At each junction, the activity has to 'choose' the thicker line.[2]

Now suppose the starting point for thinking about a particular problem, given the way it has been initially construed, is at point 'A'; and where you need to get to, the 'solution', is at point '!'. If you follow the thickness of the lines, you will see that the nature of this bit of the network is such that you can never get from A to !. You just keep going round in a circle. However, if for some reason you were to stop thinking about the problem in terms which require you to start at A, and were instead, by accident, to access the same bit of circuitry via another point, B, you would 'magically' find that now you *are* able to get from B to ! very easily. 'BING!', as Konrad Lorenz would say. You have an 'insight'. When you drop out of d-mode, and just let the mind drift around all kinds of 'irrelevant' or even 'silly' associations, you may well, by chance, find yourself thinking not about A but about B – and the recalcitrant solution suddenly becomes obvious. Thus it is, for example, that the most effective antidote to the 'tip-of-the-tongue' state is to stop *trying* to recall the word that stubbornly refuses to come to mind, and to allow yourself to drift off, or to do something else. And then, at some unpredictable moment, the word comes to you, out of the

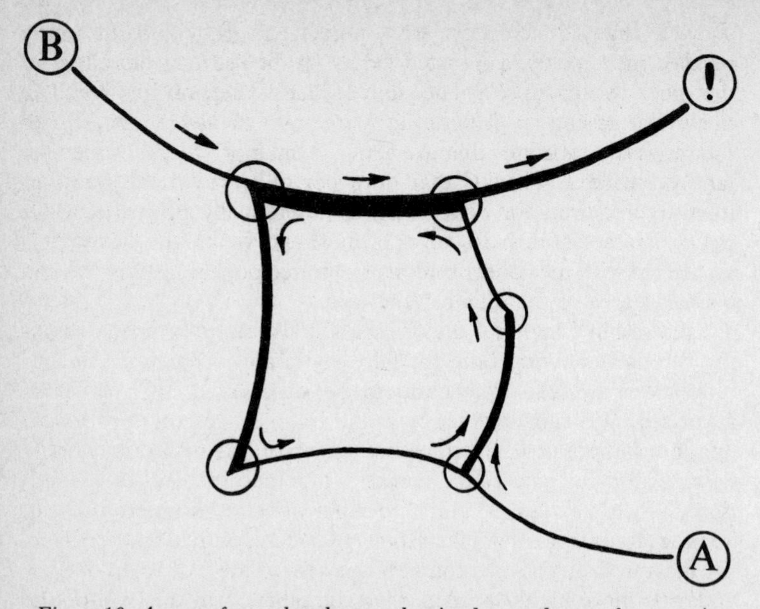

Figure 10. A map of neural pathways, showing how a change of entry-point can make an insoluble problem soluble. Activation has to follow the thicker line at each junction.

blue. The head-on assault contains the seeds of its own destruction, while the sideways approach, in which you allow the undermind to sneak up on the item you want, wins the day. That is why brainstorming and daydreaming are – as creative people have always known – effective ways of knowing: they capitalise on the brain's biochemistry.

This neural model also makes it clear why creativity favours not just a relaxed mind, but also one that is well- but not over-informed. In those parts of the neural network that represent the most familiar or routine areas of life, the continual repetition of patterns in experience may have carved out mental canyons and ravines so steep-sided that even when excitation is generally increased and inhibition relaxed, the course of activation flow will still be set. We cannot but construe the world in terms of concepts that are so engrained. However, where the brainscape is contoured enough to formulate an interesting problem, but not so deeply etched that a single approach is inescapable, then moving to the broad focus mode may well reveal novel associations.

It is widely assumed that the total amount of the brain that can be active at any one time varies only within a circumscribed range: there can be only so much activation to 'go round'. At higher levels of arousal the pool may be increased somewhat, but it is clear that if activation were allowed to proliferate unchecked, cognition would lose any sense of direction or definition at all. We might experience an entertaining psychedelic firework display of ever-expanding associations and allusions, but we would rapidly become swamped by them, unable to discriminate the useful and pertinent from the random and trivial. (This is exactly what happens in certain kinds of brain disorder. A famous case was described by the celebrated Russian neurologist A. R. Luria in his book *The Mind of a Mnemonist*.)[3] As patterns of activity move through the brain, there must be inhibitory mechanisms that 'turn off the lights behind them'.[4]

The assumption of limited resource helps to explain why thinking in words can impede non-verbal, more intuitive or imaginative kinds of cognition, and how it is possible to become clever at the expense of being wise. Some of the conceptual hollows in the brainscape are labelled; they have been given names, like 'Jane' or 'breakfast' or 'cat'. Names naturally pick out and focus attention on those features and patterns of the concept that are most familiar and essential: they tend to be associated with the nub of features at the bottom of the hollow, rather than with those that are on the slopes. Slightly fancifully, we might extend the landscape metaphor by planting a tall flagpole at the centre of such articulated concepts, at the top of which flutters a flag bearing the concept's name. This image will serve provided we remember that the 'flag' represents another set of neural patterns, to which the concept is linked, corresponding to the way the word sounds, looks, is spoken and written.

As a child learns language, the flags proliferate, and themselves become connected together into strings of linguistic bunting that begin to create a 'wordscape' that overlays the experientially based brainscape. Words can be combined to 'name' concepts that have no underlying reality, no direct conceptual referent, in that person's experience. Such verbal concepts are heavily influenced by the categories of a particular culture, and conveyed, moulded, through both formal and informal tuition. Different languages carve up the world of experience in different ways. The Inuit famously have dozens of words for 'snow'. English has no concept that even remotely resembles the Japanese *bushido*, the warrior code that combines fighting skill, considerable cruelty and aesthetic and emotional

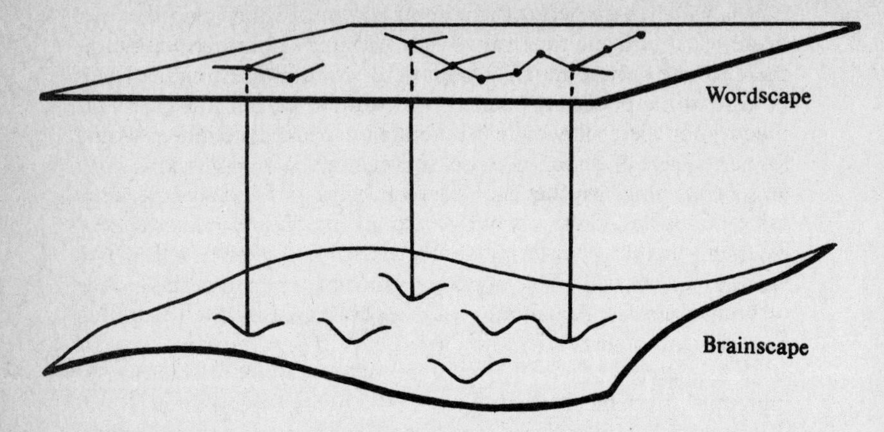

Figure 11. Brainscape and wordscape. Some concepts have no labels; and some labels are not directly underpinned by concepts.

sensitivity. The topography of each 'plane', the brainscape and the wordscape, and the relationship between them, represents an evolving compromise between the erosion of the brainscape by first-hand experience, and the dictates of a language about what segments and groupings are to be named.[5]

This model generates a brain-based account, for example, of the fact (referred to in Chapter 6) that describing faces can make them harder to recognise. Instead of simply focusing on the face as a unique whole, and allowing a rich pattern of connections between neuron clusters to become associated together in the brainscape, energy has to be put into making the face conform to general physiognomic concepts that have verbal flags. One is forced to construct a representation that is based on stereotypes ('bulbous nose', 'bushy eyebrows'); is motivated by what can be said rather than what is there; and is an accumulation of fragments rather than a holistic impression.

When a neural cluster, a concept, is activated diffusely, the intensity of activity at its focus may not be sufficient to trigger the verbal label that is attached to it. We have seen that it is possible to be aware of something, both consciously and unconsciously, without being able to retrieve its name. But when an adequate pool of excitation has concentrated at the epicentre of the concept, that activation may flow into, and activate, the representation of the name, which may, in turn, set off a chain of verbal descriptions and associations. And when verbal propositions become activated, the total reservoir of activation is depleted. The fact that the total pool of activation is limited means that activation of a portion of the wordscape must be at the expense of other possible activations and movements within the non-verbal parts of the brainscape. So the more activation that has been 'syphoned up' into the wordscape to subserve an attempt to construct conscious theories and explanations, the less is left to activate other areas of the brainscape. Activation that is focused and verbal tends to support concepts and ways of thinking that are more highly abstract, lacking broader detail and resonance.

In particular, widespread, low-intensity activation incorporates into the representation of a situation more of its personal resonances and connotations. The more the ripples of association spread out, the richer the pattern of meanings that is activated. Objects of perception or thought are imbued with greater significance, because their representation is shot through with a person's felt concerns: their hopes, fears, plans and interests. Situations therefore 'make more sense'; we know more clearly where we stand when the sensory impressions they generate are grounded and glued together by feeling. Because attention is broadly and dimly distributed, these elements of feeling may not themselves emerge into the bright light of consciousness, but nevertheless their activation ensures that perception is suffused with significance.

Felt meaning is embodied. When we sense significance deeply, it affects us physically. We do not simply understand: we are 'touched' or 'moved'. Much of the warp of significance that is woven into the fabric of sensation consists of such bodily connotations. In general, the inner as well as the outer senses contribute to the overall flow and pattern of the brain's activity, and when activity is diffused, the state of the viscera and musculature, any bodily felt *emotions, needs or threats*, are incorporated within the representation as a whole.[6]

Conversely, as the focus of activation tightens, the image of the world that is created becomes more abstract, more intellectualised, and less rich in meaning and feeling. David Gelernter, who has reviewed much of the evidence for the link between focus and emotion, concludes:

> As we inch upwards, gradually raising or tightening our focus . . . thinking starts unmistakeably to grow numb. We are less and less able to *feel* our recollections; we merely witness them

> . . . Thought loses its vividness . . . In the end we are left to the cold comfort of logic alone to peg together a powerful and penetrating – [but] numb and pale – thought-stream and drive it forward.

The trains of thought that may be stimulated in the wordscape are also likely to be more rigid, more stereotypical and more defined by the conventions of the linguistic culture at large, than the patterns of the brainscape. Thus it may well be harder – as many creative people have argued – to be original in propositions than in intuitions, or to unearth and question cultural assumptions that are embodied in the very way the wordscape is constructed.

One crucial aspect of the functioning of the brain remains to be explored: how and where and why does it create consciousness? Even to pose the question in this way makes one important assumption: that consciousness *is* a product of the brain, rather than, for example, a universal property of all matter, or a signal from elsewhere that is picked up by the brain in the same way that a broadcasting channel is detected and transformed by a television receiver. Both of these views have lengthy histories in philosophy and religion, but I shall join the neuroscientific consensus in seeing consciousness as a correlate of certain kinds of activity that occur only in nervous systems of a particular kind and degree of complexity. The prima facie evidence for this starting point is, after all, overwhelming. We know through direct experience that members (or, if we are being absolutely precise, at least one member) of one complex species – *homo sapiens* – possess consciousness, whereas the idea that amoebas, daffodils or pebbles are conscious is at best a conjecture or a projection. And we know that it is damage to the central nervous system – rather than to the liver or the lungs, for example – that results in alterations to or even loss of consciousness.

But if consciousness is a property of brains, we can be sure that it is not a property of individual neurons. We see and think in terms of concepts and images that are associated with large groups of interwoven neurons, not single cells. And we may also discard the idea that consciousness can be localised within some particular area or structure of the brain. Despite many attempts to find such a specific anatomical substrate for consciousness, none has been discovered. Nor should we really expect it to, given that we know that a single neural cluster is itself distributed widely throughout the brain. Descartes thought that the pineal gland, situated right at the centre of the brain, was the 'seat of the soul' and, as Daniel Dennett describes it, the door into the mental cinema where consciousness is projected on to a screen. But we now know without doubt that there is no localised 'headquarters' in the brain to which all inputs are referred, and from which all orders are issued.[7]

We have to think of consciousness, therefore, as associated with *states* of the nervous system rather than places. So our question has to be: what are the conditions which are necessary and/or sufficient for the brain to make its activities conscious? The short answer is that nobody knows for sure. Finding an answer to this question is the 'holy grail' of neuro- and cognitive science at present – as it has been, recurrently, for philosophers and theologians over the centuries. But there are some clues. First, consciousness is associated with *intensity*. The strength or the concentration of activation in a cluster of neurons seems to play a role. Signals that are loud or bright or shocking grab our conscious attention, and stimuli which are being processed unconsciously can be 'boosted' into consciousness if their magnitudes are increased.

For example, even blindsight patients can become directly conscious of activity in the blind field if a stimulus suddenly gets brighter or starts to move faster. In another neurological disorder called prosopagnosia, people are specifically unable to recognise faces that they do in fact know well. Show them Princess Diana and they will not know that they know the face, nor be able to put a name to it. But if these patients have previously been shown a picture of Prince Charles – which they have not been able to identify either – the chances of their consciously recognising Diana are increased. Even though Charles did not himself exceed the threshold of consciousness, his picture has been unconsciously recognised, and this is sufficient to send some activation to prime the underlying Diana network. This, added to the activation from Diana's picture, may be strong enough for the image to reach consciousness.[8]

But intensity alone cannot account for consciousness. Even very strong stimuli can, with time, be ignored. And there is evidence that some neurons, paradoxically, are *more* strongly stimulated by subliminal stimuli than by conscious ones. There are cells in the visual area of the brain, for example, that give a more vigorous response to a light that is shone in the eye of a fully anaesthetised animal than one that is wide awake.[9]

A more crucial condition for consciousness seems to be the *persistence* of neural activity. Benjamin Libet at the University of California at San Francisco has found, by directly stimulating the part of the brain that is responsible for the sense of touch, that even quite strong stimuli must reverberate in the brain for a minimum period of about half a second before they become conscious, whereas reactions to unconscious processing can occur much faster, and with much briefer stimuli.[10] If there is such a minimum time condition for consciousness, then it is possible that people who are asked to move from an unconscious to a conscious mode of responding might show not a smooth increase in their reaction time, but a jump as they move from one mode to the other. Sure enough, the time it takes for people to push a button, as quickly as they can, in response to the onset of a light, is about 200 milliseconds. But if they are asked to slow down their responding by the tiniest possible amount, there is a quantum jump of about half a second, giving total response times of nearly three-quarters of a second. It is as if there is no halfway house: either you are responding instinctively, or you have to wait for consciousness to develop (like a photograph) and then respond.[11]

There are a number of different reasons why the neural effect of a stimulus might last long enough for consciousness to develop. One, obviously, is the duration of the stimulus event itself. But strong stimuli may become conscious, even if they are objectively too short, simply because intensity causes neural activity to reverberate for a longer time – just as the sound of a well-struck gong hangs in the air. (In this way we can subsume the 'intensity' condition under the 'persistence' condition.) And such reverberation may come about not just as a result of the strength of a stimulus. Semir Zeki at University College London has proposed, in the case of vision, that two different areas of the visual cortex must be able to 'sustain a dialogue' if consciousness is to occur. Francis Crick and Christof Koch at the Salk Institute in California have made a somewhat similar suggestion: that what is required to subserve consciousness is a reverberatory loop linking the thalamus in the midbrain to the neocortex.

A third possibility, and one which ties in with some of the experimental evidence we looked at in Chapter 8, is that it is the involvement of the 'self' that extends neural activity to the point of consciousness. We have just seen that asking people to respond self-consciously, rather than instinctively, seems to bump them into a qualitatively different, slower mode of processing. And we have also seen the reverse effect: if people are asked to respond very fast to weak signals – to jump from a careful reliance on the cautious criteria of consciousness to a fast, unchecked way of responding –

they give correct answers which consciousness, when it catches up, countermands. From this point of view consciousness may be *sui generis* self-consciousness.[12] We become conscious of stimuli (except when they are in themselves strong or persistent) because they are being referred to a special part of the neural network which corresponds to our self-image, to see if they fit comfortably with the sense of who we are, and with the ongoing life story in which we see ourselves as taking part. This checking process takes time, and thus in itself tends to fulfil the conditions required for neural activity to generate consciousness (though other censoring processes may quickly come into play – as we saw with the phenomenon of 'perceptual defence' – if it turns out that the information is judged, on this analysis, to be threatening or uncongenial).[13]

If a stimulus is either too short or too weak, or does not fit with my model of who I am and what is going on, it can still play its part in determining how the brain is reacting to events, but it will do so covertly rather than explicitly. As John Kihlstrom, one of the leading researchers on the cognitive unconscious, sums it up:

> When a link is made between the mental representation of self and the mental representation of some object or event, then the percept, memory or thought enters into consciousness; when this link fails to be made, it does not. Nevertheless, unconscious percepts and memories, images, feelings and the like can still influence ongoing experience, thought and action . . .[14]

This association between the self and consciousness, and the idea that consciousness demands a time-consuming resonance or reverberation between different circuits of the brain, raises the intriguing possibility that areas of the brain might be getting on with their business at an unconscious level, without bothering to wait for consciousness to develop. A pool of neural activation may split into two, one part resonating with the self, and thus subserving the emergence of conscious awareness, while the other carries on with further processing such as planning a response. This makes sense, particularly if time is short. It may pay you to continue with the preparations for building an extension on to your house at the same time as you are waiting for formal 'planning permission' – on the assumption that the permission will be forthcoming in the end. If permission is in the event refused, you can abort the plan before you have started the actual process of construction. Provided you have not physically 'jumped the gun', nothing, except the planning time, is lost. If the brain were capable of operating in this dual-track manner, we would have to rethink the function of consciousness. Far from being the instigator of action, the *source* of orders and decisions, consciousness could, at least under some conditions, simply be receiving notification of what was in fact being decided elsewhere.

Another study by Ben Libet has demonstrated that this bifurcation of the mind does occur.[15] He asked people to hold out a hand and, whenever they felt like it, to flex one of their fingers. While they were doing this simple task, he recorded three points in time. Firstly, by virtue of electrodes attached to the person's head, he was able to pinpoint the moment at which the precursors of the action were discernible in the brain's patterns of electrical activity (the EEG). Secondly, Libet asked people to indicate, by registering the position of a spot on a rotating clock in front of them, the moment at which they were first aware of the intention to make the movement. And finally, by recording activity in the muscles of the finger, he was able to note when the physical movement itself began. He found that the voluntary action began to develop in the brain about 350 milliseconds – a third of a second – *before* the appearance of the conscious intention, which occurred, in turn, some 200 milliseconds before the start of the action itself. These results indicate clearly that it is the unconscious brain which decides what to do, and when; and that what we experience as an intention is merely a *post hoc* confirmation of what has already been set in motion. Consciousness receives a kind of corollary 'despatch note', and then presents this as if it were the original order.

'Will', or even 'free will', on this evidence, seems to belong to the brain, rather than to consciousness. But this does not mean that consciousness is left without any function at all. If conscious awareness is associated with the process of checking a situation for concealed threats to self, it may also be instrumental in inhibiting actions and experiences that are adjudged to be risky, rather than in routine instigation and construction. The detection of some irregularity or threat, real or imagined, may result in a veto that can block the execution of the evolving plan before the 'point of no return'. As psychologist Richard Gregory has speculated, it may be that we – that is, consciousness – do not have 'free will', but we do have 'free won't'.

This brings us to the question: what is consciousness for? We are most conscious of those things which might threaten us (except when the feeling of threat is so great, so threatening in itself, that the very experience is inhibited, as in hysterical blindness, traumatic amnesia or psychopathological forms of repression). Where the brain's initial unconscious diagnosis declares the situation to be safe and familiar, there is no need to dwell on it – no need to refer it to the self for further tests. The flow of activation moves on too quickly for the persistence condition for consciousness to be met. But where there is some initial doubt, then the flow of activation is arrested and the predicament is allowed to resonate with the priorities and checks of the self, so that further data may be collected, or a wider pool of associations activated. If this results in the 'all clear', action can proceed unhindered. If a threat is uncovered, then censorship and self-control can save the day. Consciousness is for self-protection.

From this perspective, conscious awareness emerges as fundamentally quizzical and questing. We become conscious of that which is being actively probed for its significance. As many cognitive scientists have recently argued, focused, conscious awareness, the spotlight of the mind, is primarily associated with states of disruption or emergency, and with the activities of investigation, detection and resolution that follow.[16] Some segment of the world poses a puzzle, and high levels of activation concentrate in the corresponding areas of the neural network, so that the nature of the predicament – 'What *is* it, out there (or in here)?' – can be discerned more fully, and an appropriate response discovered. In an emergency, or a state of utter absorption, the proportion of the brain's (limited) resources that are appropriated by this enquiry may be so great that other competing activities are temporarily shut down – we freeze, we stop breathing. Consciousness, it has been suggested, originally made its evolutionary appearance in the context of this focused, arrested response to threat or uncertainty.

This view of the mind is very different from the commonsense one, which sees consciousness as the executive boardroom of the mind, and as the theatre in which 'reality' is displayed. Specifically, the evidence seems to suggest two rather radical conclusions. First, common sense tells us that consciousness is what we can trust: that the world is as it appears to be. But brain research indicates that consciousness manifests not what is certain, but what is in question. Focused consciousness is associated with those aspects of the mind's activity *which are currently being treated as problematic.* Whatever occupies the centre of conscious attention is there precisely because its meaning, its significance, its interpretation, is in doubt. Through dwelling on something we may be led to a richer, 'truer' understanding of it, but that is the result of being conscious, not the prerequisite.

The second conclusion is that consciousness *per se* does not actually *do* anything. Consciousness accompanies, and is therefore symptomatic of, a particular mode of operation of the brain-mind *as a whole*: one in which ongoing action is arrested, careful attention to the probable source of the interruption is being paid, all the sub-systems are listening carefully for new information, priorities are being revised, and new plans being laid. These are the circumstances under which the brain-mind 'generates' conscious awareness. The intense pooling and resonating of activation creates the conditions in which consciousness appears. Consciousness accompanies a very particular and very useful mode of mind, though it does not itself possess any executive responsibility. Although I earlier suggested that we, qua consciousness, may possess 'free won't', even this formulation credits consciousness with power that it does not intrinsically possess; for the process of vetoing action or editing experience is itself carried out by the brain. Even the 'self' turns out to be just one sub-system among many within the overall neural economy of the brain – the system that defines what is to count as 'threat' or 'desire', and which examines experience through these filters.

The idea that the brain, unsupervised by the conscious intellect, does smart things on its own, and that consciousness *per se* does not carry out any cognitive function, can be discomfiting, because it seems to leave 'us' at times with nothing to do. Yet this disconcerting feeling of redundancy may be the price we have to pay if we wish greater access to the slow ways of knowing. Certainly there are many neuroscientists who firmly believe that the physical brain is *all* we need to account for human intelligence. One of the clearest spokespeople for this point of view is the self-styled 'neurophilosopher' Patricia Churchland, who has written:

> The cardinal principle for the [neuroscientist] is that . . . there is no little person in the brain who 'sees' an inner television screen, 'hears' an inner voice . . . weighs reasons, decides actions and so forth. There are just neurons and their connections. When a person sees, it is because neurons, individually blind and individually stupid neurons, are collectively orchestrated in the appropriate manner . . . In a relaxed mood we still understand perceiving, thinking, control and so forth, on the model of a self – a clever self – that does the perceiving, thinking and controlling. It takes effort to remember that the cleverness of the brain is explained not by the cleverness of a *self* but by the functioning of the neuronal machine that is the

> brain . . . In one's own case, of course, it seems quite shocking that one's cleverness should be the outcome of well-orchestrated stupidity.[17]

Perhaps any discomfort that these ideas cause is merely a product of having identified ourselves too closely with the habits and values of d-mode, and that all that is required to dispel the unease is an expansion of this shrunken definition of intelligence to reincorporate the brain.

CHAPTER 11

Paying Attention

One day a man of the people said to Zen master Ikkyu: 'Master, will you please write for me some maxims of the highest wisdom?' Ikkyu immediately took his brush and wrote the word 'Attention'. 'Is that all?' asked the man. 'Will you not add something more?' Ikkyu then wrote twice running 'Attention. Attention'. 'Well,' remarked the man rather irritably, 'I really don't see much depth or subtlety in what you have just written.' Then Ikkyu wrote the same word three times running: 'Attention. Attention. Attention'. Half-angered, the man demanded: 'What does that word "Attention" mean anyway?' And Ikkyu answered gently: 'Attention means attention.'

Philip Kapleau Roshi

In d-mode, perception is diagnostic. Its role is to sample the information that is arriving through the senses until it can recognise, categorise and label what is 'out there' – a 'traffic jam', a 'politician' – or 'in here' – 'sadness', a 'headache'. Once perception has come up with its diagnoses, its job is done, and interest shifts downstream to what can be inferred, and done, about the situation thus described. If the snap diagnosis is accurate and adequate, thought builds on a firm foundation. But there is always a risk that such a skimpy approach to perception may neglect information that does not, on first sight, *seem* to be significant, but which, had attention been less precipitate, might have revealed its relevance and its worth. D-mode determines the way in which attention is to be deployed, and it is not always the best way. If we get stuck in d-mode's particular way of attending, we may prematurely and unwittingly discard just what we need. Sometimes a slower, more meticulous approach to perception can lead to a richer mental image of what is happening, and hence to a better way of knowing. If we are to know as well as we can, we sometimes need to switch from the high-speed scanning of d-mode into a contemplative perceptual stance in which the world is allowed to speak more fully for itself. This chapter explores four different ways of paying attention, or 'slow seeing': detection, focusing on inner states, poetic sensibility, and mindfulness.

The habit of attending closely and patiently to the evidence, even – sometimes especially – to tiny, insignificant-looking shreds of evidence, is characteristic of skilled practitioners of a variety of arts, crafts and professions, prototypically the hunter. From a bent twig, a feather or a piece of dried excrement the expert hunter can recreate an animal, its age and state of health; and he does so in an apparently leisurely fashion in which these scraps of information are allowed to resonate, largely unconsciously, with his mental stock of lore and experience. You can't rush a tracker. Each detail, slowly attended to, is allowed to form a nucleus, an epicentre in the brain, around which associations and connotations gradually accrete and meld, if they will, into a rich, coherent picture of the animal and its passage. As Carlo Ginzburg, author of a fascinating essay on 'Clues', has surmised, the hunter squatting on the ground, studying the spoor of his quarry, may be engaged in the oldest act in the intellectual history of the human race.[1] Many other feats of vernacular connoisseurship – telling an ailing horse by the condition of its hocks, an impending storm by a change in the wind, a run of salmon by a scarcely perceptible ripple on the river, a hostile intent by a subtle narrowing of the eyes – are of the same kind. Each is an act of high intelligence, bringing to bear on the present a complex body of past knowledge, and accomplished by the eye, with little if any assistance from deliberate thought.

In this process of attentive resonance, knowledge does not become the object of explicit thought; rather it implicitly dissolves itself in a gathering sense of the situation as a whole. There is an apocryphal story of a venerable factory boiler that broke down one day, and of the old man who was called to fix it. He wandered around among its convoluted pipework, humming quietly to himself and occasionally putting his ear to a valve or a joint, and then pulled a hammer out of his toolbag and banged hard on one small obscure corner. The boiler heaved a deep sigh and rumbled into life again. The old man sent in a bill for £300, which the manager thought excessive, so he sent it back with a request that it be itemised. When it came back, the old man had written:

for tapping with hammer: 50p
for knowing where to tap: £299.50p.

Similarly, the painter J. M. Whistler, at the trial of John Ruskin, was asked by the judge how he dared ask £350 for a 'Nocturne' that had taken him only a few hours to paint. Whistler replied that the fee was not for the painting, but for 'the knowledge of a lifetime'.

In the late nineteenth century, three new professions came into being that explicitly relied on the ability to read clues: the authentication of artworks, police detection, and psychoanalysis. In the mid-1870s, Giovanni Morelli developed a method for discriminating original paintings from copies and fakes, based not on overall composition or draughtsmanship but on the execution of such tiny details as earlobes and fingernails. He argued that it was precisely in these unimportant details, when both the 'master' and the copyist were 'off guard', that differences of technique would manifest themselves most clearly. As with a casual signature, rather than a self-conscious script, it was in these inadvertent trivia that personality would reveal itself – but only to the eye which understood this to be the case. Like the hunter, one had to be alert to the presence of meaning in the scraps and marginalia.

Morelli directly influenced the development of the 'science' of detection, which was to be dramatised by the emerging writers of detective fiction, such as Gaboriau in France in the late 1870s, and, a little later, most famously, by Sir Arthur Conan Doyle in his Sherlock Holmes stories. Gaboriau, in one of his 'Monsieur Lecoq' adventures, contrasts the novel approach of the detective Lecoq with the 'antiquated practice' of the old policeman Gevrol, 'who stops at appearances, and therefore does not succeed in seeing anything'.[2] While in the Sherlock Holmes story called 'The Cardboard Box', which begins with the mysterious arrival, at the home of 'an innocent maiden lady', of a box containing two severed ears, Holmes literally 'morellises'. Dr Watson reports: 'Holmes paused, and I was surprised, on glancing round, to see that he was staring with singular intentness at the lady's profile.' And Holmes later explains:

> You are aware, Watson, that there is no part of the body which varies so much as the human ear . . . I had, therefore, examined the ears in the box with the eyes of an expert, and had carefully noted their anatomical peculiarities. Imagine my surprise then, when, on looking at Miss Cushing, I perceived that her ear corresponded exactly with the female ear which I had just inspected . . . I saw at once the enormous importance of the

> observation. It was evident that the victim was a blood relation, and probably a very close one.[3]

Sigmund Freud too was influenced, in his developing formulation of the psychoanalytic method, by Morelli, and quite possibly by Conan Doyle as well. Freud is recorded as speaking of his fascination with the Sherlock Holmes stories to one of his patients (the so-called 'wolf-man'). Certainly he had become intrigued by the techniques of Morelli at least ten years before he began to develop his ideas about psychoanalysis in print. In a retrospective essay, 'The Moses of Michelangelo', published in 1914, Freud writes of this influence thus:

> Long before I had any opportunity of hearing about psycho-analysis, I learnt that a Russian art-connoisseur, Ivan Lermolieff [a pseudonym of Morelli's], [was] showing how to distinguish copies from originals . . . by insisting that attention should be diverted from the general impression and main features of a picture, and he laid stress on the significance of minor details . . . which every artist executes in his own characteristic way . . . It seems to me that his method of inquiry is closely related to the technique of psycho-analysis. It, too, is accustomed to *divine secret and concealed things from unconsidered or unnoticed details, from the rubbish heap, as it were, of our observations.*[4] (Emphasis added)

It is interesting to observe, in this context, the changing approach to medical diagnosis over the course of the last two hundred years. The process of detection and identification of disease these days is often devoid of this leisurely resonance of attentive observation with the working knowledge of a lifetime's experience. The modern general practitioner makes a succession of snap decisions as to either the nature of the disorder with which she is confronted, or what further objective, 'scientific' tests to order. She is now so rushed, and so enchanted (as we all are) by technology, and technological ways of thinking, that she generally prefers to trust a read-out from a machine over a considered clinical judgement. An instrument gives us 'real knowledge' about the patient, whereas the poor doctor on her own can offer nothing more substantial than an 'opinion'. Reliance on informed intuition seems increasingly 'subjective', risky and old-fashioned. As medical historian Stanley Reiser says:

> Without realising what has happened, the physician in the last two centuries has gradually relinquished his unsatisfactory

> attachment to subjective evidence . . . only to substitute a devotion to technological evidence . . . He has thus exchanged one partial view of disease for another. As the physician makes greater use of the technology of diagnosis, he perceives his patient more and more indirectly through a screen of machines and specialists; he also relinquishes control over more and more of the diagnostic process. These circumstances tend to estrange him from his patient *and from his own judgement.*[5] (Emphasis added)

Yet throughout the history of medicine, the doctor has functioned more like the tracker or the detective than a technician. And even today there are striking examples of this attentive, resonant intuition at work. There is the much-retailed account of the day the Dalai Lama's personal physician, Yeshi Dhonden, visited Yale Medical School, for example. He gave a demonstration to the assembled group of sceptical Western doctors of traditional Tibetan medical diagnosis by examining a woman patient with an undisclosed illness. On approaching the woman's bed, Yeshi Dhonden asked her no questions, but simply gazed at her for a minute or so before taking her hand and feeling for her pulse. Richard Selzer was one of the physicians present:

> In a moment he has found the spot, and *for the next half-hour* he remains thus, suspended above the patient like some exotic golden bird with folded wings, holding the pulse of the woman beneath his fingers, cradling her hand in his. All the power of the man seems to have been drawn down into this one purpose. It is palpation of the pulse raised to the state of ritual . . . his fingertips receiving the voice of her sick body through the rhythm and throb she offers at her wrist. All at once I am envious – not of him, not of Yeshi Dhonden for his gift of beauty and holiness, but of her. I want to be held like that, touched so, *received.* And I know that I, who have palpated a hundred thousand pulses, have not felt a single one.

Finally Yeshi Dhonden laid the woman's hand down. He turned to a bowl containing a sample of her urine, stirred it vigorously, and inhaled the odour deeply three times. His examination was over. He had still not uttered a single word. His diagnosis, whatever it was, would be based solely on his protracted attention to the appearance, the feel and the smell of the woman's sick body. Back in the conference room Yeshi Dhonden, through his young interpreter, delivered his verdict in curiously poetic terms. 'Between the chambers of the heart, long, long before she was born, a wind had come and blown open a deep gate that must never be opened. Through it charge the full waters of her river, as the mountain stream cascades in the springtime, battering, knocking loose the land, and flooding her breath.' Finally the woman's consultant disclosed his diagnosis: 'congenital heart disease: interventricular septal defect with resultant heart failure'. Unless he was very lucky, or had been secretly primed, we may conclude, with the originally sceptical Selzer, that Yeshi Dhonden was 'listening to the sounds of the body to which the rest of us are deaf'. Having stilled his mind through the practice of meditation, he looks, listens, feels and smells *without thinking*, without trying to make any sense, allowing all his sensory impressions to seep at their own speed into the furthest corners of his vast, largely inarticulate storehouse of knowledge, and to deliver back to him, in consciousness, images and figures that make sense of the whole.[6]

This kind of detection comes into its own under certain conditions. It needs a problem that can be clearly formulated – how long since the horses passed?; who planted the bomb?; what is causing the fever? – but to which the answer is not obvious. It requires 'clues': pieces of information whose significance, or even presence, is not immediately apparent. It works with a mind that has a rich database of potentially relevant information, much of which is tacit or experiential rather than articulated. And this kind of detection requires a particular mental mode in which details can be dwelt upon, at first without knowing what their meaning may be, so that slow ripples of activation in the brain may uncover any significant connections there may be. Without this patient rumination, the clue, the problem and the database will not come into the fruitful conjunction that reveals the ways in which they are related.

The successful detective trains her awareness on the outside world, in order to find meaning in the minutiae of experience. The second fruitful way of paying attention is similar, except awareness is now directed inward, towards the subtle activities and promptings of one's own body. The ability to 'listen to the body' is very useful in gaining insight into a whole variety of personal puzzles and predicaments. This ability has been dubbed *focusing* by the American psychotherapist Eugene Gendlin. Back in the 1960s, Gendlin and colleagues at the University of Chicago were involved in large-scale research project designed to discover why it was that some people undergoing psychotherapy made good progress while others did not, no matter who the therapist was or what she did. After analysing thousands of hours of tape-recorded sessions, Gendlin uncovered the magic ingredient, which could be picked up even in the first one or two sessions, and which would predict whether the client would make progress or not. It was not anything to do with the school or the technique of the therapist, nor, apparently, with the content of what was talked about. It was the clients' spontaneous tendency to relate to their experience in a certain way. If they did, they would make progress; if they did not, they wouldn't.[7]

The successful clients were those who spontaneously tended to stop talking from time to time; to cease deliberately thinking, analysing, explaining and theorising, and to sit silently while, it seemed, they paid attention to an internal process that could not yet be clearly articulated. They were listening to something inside themselves that they did not yet have words for. They acted as if they were waiting for something rather nebulous to take form, and groping for exactly the right way of expressing it. Often this period of silent receptivity would last for around thirty seconds; sometimes much longer. And when they did speak, struggling to give voice to what it was they had dimly sensed, they spoke as though their dawning understanding was new, fresh and tentative – quite different from the tired old recitation of grievance or guilt which frequently preceded it.

Gendlin called this hazy shadow which they were attending to, and allowing slowly to come to fruition, a *felt sense*, and it was quite different both from a string of thoughts and from the experience of a particular emotion or feeling. It seemed to be the inner ground out of which thoughts, images and feelings would emerge if they were given time and unpremeditated attention. It appeared that many people lacked the ability, and perhaps the patience, to allow things to unfold in this way. Instead they would, in their haste for an answer, pre-empt this process of evolution, creating a depiction of the problem which told them nothing new, and which gave no sense of progress or relief.

Gendlin discovered that the felt sense will form not in the head, but in the centre of the body, somewhere between the throat and the stomach. The awareness is *physical* and when it has been allowed to form, has been heard, and accurately captured in a phrase or an image, there is a corresponding physical sense of release and relaxation. It is as if some inarticulate part of the person, almost like a distressed child, feels understood, and has responded with a sigh of relief: 'Yes. That's exactly how it is. You understand. Thank you.' When this 'felt shift' happens, then the previous feeling of blockage eases, and by going back again patiently to the felt sense, people find that it is ready to tell them something further; to unfold a little more.

In focusing one takes an issue to consider, asks oneself 'What is this whole thing about?', and then *shuts up*. Over the course of half a minute or so, by holding awareness in the body, a physical sense of 'the whole thing' begins to form in a way that, at first, is unsegmented, and therefore inarticulable. The normal d-mode-dominated tendency to leap to conclusions, to construct a clear and plausible narrative as quickly as possible, is reversed. Answers from d-mode, which tend to come quickly and with a veneer of 'this-is-obviously-the-way-it-is' certainty, are ignored.[8] You know you are doing 'focusing' right, according to Gendlin, when you are not sure if you are doing it right – because you cannot yet *say* what is there. 'The body is wiser than all our concepts', he says, 'for it totals them all and much more. It totals all the circumstances we sense. We get this totalling if we let a felt sense form in inward space.[9]

Because this 'way of knowing' had not previously been identified as one of the main active ingredients in successful therapy, many therapists were unaware of the need to cultivate the client's ability in this regard. Yet, Gendlin discovered, once it was recognised it could be 'taught' quite directly. Anybody, with practice, could learn how to do it, and could benefit from it, not just in dealing with the kinds of problem that took people into therapy in the first place, but in a whole variety of situations in everyday life. To begin with, focusing feels strange, because it really is a different way of knowing from the one with which people are most familiar. As with a medical student learning to read X-rays, it takes time to 'see' what is there, and to stabilise these unfamiliar, shadowy objects of attention. But the tentative, exploratory 'feel' of focusing soon becomes unmistakeable. In one session in which I took part, the focuser said: 'I feel kind of scared, but I don't know what of. Inside it's like an animal that's totally alert, ears pricked . . . It's like something's coming, and some part of me has picked it up and is getting ready for it, but "I" don't know what it is yet.' It is this sense of the imminence of meaning not yet revealed that characterises focusing. The fruit of the felt sense is often an image or an evocative phrase, rather than a fully fledged story – such as the image quoted above of a startled animal, sensing danger, or the unknown, but not yet able to identify it. The first form that the emerging meaning takes is often poetic or symbolic, rather than literal and transparent.

Focusing is not, of course, a new discovery (though turning it into a technology certainly fits with the Promethean spirit of the age). It is very akin, for example, to the Japanese concept of *kufū*, which D. T. Suzuki in *Zen and Japanese Culture* describes as:

> not just thinking with the head, but the state when the whole body is involved in and applied to the solving of a problem . . . It is the intellect that raises a question, but it is not the intellect that answers it . . . The Japanese often talk about 'asking the abdomen', or 'thinking with the abdomen', or 'seeing or hearing with the abdomen'. The abdomen, which includes the whole system of viscera, symbolises the totality of one's personality . . . Psychologically speaking, [*kufū*] is to bring out what is stored in the unconscious, and let it work itself out quite independently of any kind of interfering consciousness . . . One may say, this is literally groping in the dark, there is nothing definite indicated, we are entirely lost in the maze.[10]

It may also have been Gendlin's 'felt sense' which was referred to as *thymos* by the classical Greeks. Located in the *phrenes*, again the central part of the body – lungs, diaphragm, abdomen – *thymos* is that part of a person which 'advises him on his course of action, it puts words into his mouth . . . He can converse with it, or with his "heart" or his "belly", almost as man to man . . . For Homeric man the *thymos* tends not to be felt as part of the self: it commonly appears as an independent inner voice.'[11] It appears that, in other cultures and other times, 'thinking with the abdomen' was a routine and familiar way of knowing. It is only in our contemporary European d-mode culture, dominated by the idea that thinking is the quick, conscious, controlled, cerebral manipulation of information, that the ability to think with the body has to be isolated, repackaged and taught as a novel kind of skill.

With focusing one has, as with detection, a predetermined agenda – a problem to solve or clarify – and the process of dwelling on the details is therefore circumscribed and channelled by a purpose. There is openness and patience, but there is also a background monitoring of progress and relevance. However, the third way of paying attention I want to consider, *poetic sensibility*, has the ability to reset or create our agenda; to uncover issues and reveal concerns, perhaps in unexpected quarters, or surprising ways. By allowing ourselves to become absorbed in some present experience without any sense of seeking or grasping at all, we can be reminded of aspects of life that may have been eclipsed by more urgent business, and of ways of knowing and seeing that are, perhaps, more intimate and less egocentric. As we gaze out to sea or up at a cloudless sky, listen to the sound of goat-bells across a valley or to a Beethoven quartet, we may sense something that lies beyond the preoccupations of daily life. We feel perhaps a kind of obscure wistfulness, a bitter-sweet nostalgia for some more natural, more simple facet of our own nature that has been neglected.[12]

Returning home from a day in the country, people commonly feel calmer, more whole, more balanced. We may not have understood anything, not arrived at any insights or answers, yet we may feel somehow transformed, as if something healing or important has been intimated, but not revealed. In some moods it is possible to gain glimpses of what seems to be knowledge or truth of a sort – of a rather deep sort, perhaps – which is *not* an answer to a consciously held question; and which cannot be articulated clearly, literally, without losing precisely that quality which seems to make it most valuable. There is a kind of knowing which is *essentially* indirect, sideways, allusive and symbolic; which hints and evokes, touches and moves, in ways that resist explication. And it is accessed not through earnest manipulation of abstraction, but through leisurely contemplation of the particular.

When we lose ourselves in the present, we do just that: lose our selves. As the linguist and philosopher Ernst Cassirer put it, the mind 'comes to rest in the immediate experience; the sensible present is so great that everything else dwindles before it. For a person whose apprehension is under the spell of this attitude, the immediate context commands his interest so completely that nothing else can exist beside and apart from it. The ego is spending all its energy in this single object, lives in it, loses itself in it.'[13] One slips away from self-concern and preoccupation into the sheer presence of the thing, the scene, the sound itself, until, as Keats said:

> Thou, silent form, dost tease us out of thought
> As doth eternity . . .

The ego, or the 'self', is essentially a network of preoccupations: a set of priorities that must be attended to in the interests of our survival, our wellbeing, or even our comfort. When the ego is in control of the mind, we act, perceive and think as if a wide variety of things – reputation, status, style, knowledgeability – mattered vitally, and as if their antitheses – unpopularity, ignorance and so on – constituted dire threats. When we are lost in the present, these conditioned longings fall away, and anxious striving may be replaced by a refreshing sense of peaceful belonging. Unskewed by hope or fear, perception is free simply to register what is there. As Hermann Hesse wrote in his essay 'Concerning the soul' in 1917: 'The eye of desire dirties and distorts. Only when we desire nothing, only when our gaze becomes pure contemplation, does the soul of things (which is beauty) open itself to us.'

By its very nature, this more dispassionate, yet more intimate, way of knowing cannot be brought about by an effort of will. It arises, if it does at all, spontaneously. The experience is like that of seeing the three-dimensional form in a 'Magic Eye' image. If you look intently at such an image with the normal high-focus gaze, scanning it for its 'meaning', all you will see, for as long as you look, is a flat field of squiggly shapes. You see plenty of detail, but it does not cohere. However, if you give up 'trying to see what's there', relax your eyes so that they gaze softly *through* the image, and stay for a while in this state of patient incomprehension, then the details begin to dissolve and melt into one another, and a new kind of seeing spontaneously emerges, one which reveals the 'hidden depths' in the picture. There is no doubt when this revelation has occurred: it has a visceral impact which cannot be forced or feigned – just as the 'getting' of a joke is a spontaneous, bodily occurrence that cannot be engineered. Someone who 'thinks' they see the image, like someone who 'understands' a joke, simply has not got it.

Though poetic sensibility cannot be commanded, it can, as with the three-dimensional visual image, be encouraged. One can make oneself prone to it by cultivating the ability to wait – to remain attentive in the face of incomprehension – which Keats famously referred to as 'negative capability': 'when a man is capable of being in uncertainties, mysteries, doubts, without any irritable reaching after fact and reason.' To wait in this way requires a kind of inner security; the confidence that one may lose clarity and control without losing one's self. Keats's description of negative capability came in a letter to his brothers, following an evening spent in discussion with his friend Charles Dilke – a man who, as Keats put it, could not 'feel he had a personal identity unless he had made up his mind about everything'; and who would 'never come at a truth so long as he lives; because he is always trying at it'.[14]

The domination of culture and education by d-mode seems to have created a whole society of Charles Dilkes: to have estranged people from a way of knowing that is, perhaps, part of their cognitive and aesthetic birthright. It certainly appears as if children may have more ready access to poetic sensibility than adults. Young children have been found to be very 'poetic' in their way of knowing in at least one respect: they are much better than older children and adults at producing and using metaphors. Psychologists Howard Gardner and Ellen Winner have found that three- and four-year-old children produce many more appropriate metaphors for a situation than do seven- and eleven-year-olds, and all children are much more fluent users and creators of spontaneous metaphor than college students.[15] And Wordsworth, in his 'Ode to Immortality', famously bemoans the loss of his childhood ways of knowing.

> There was a time when meadow, grove and stream,
> The earth, and every common sight,
> To me did seem
> Apparelled in celestial light,
> The glory and the freshness of a dream.
> It is not now as it hath been of yore; –
> Turn whereso'er I may,
> By night or day,
> The things which I have seen I now can see no more.

It may well have been the child's ability to be lost in the present that prompted the following exchange:

> 'Come along!' the nurse said to Félicité de la Mennais, eight years old, 'you have looked long enough at those waves and everyone is going away'. The answer: 'ils regardent ce que je regarde, mais ils ne voient pas ce que je vois', was no brag, but merely a plea to stay on.[16]

Though it is often lost by the time one reaches adulthood, the knack of absorption can be recaptured. One can cultivate the requisite attitude of receptivity, of allowing oneself to become quietly immersed in things – and then to wait and see. As Jacques Maritain, author of the monumental *Creative Intuition in Art and Poetry*, has said of 'poetic intuition':

> It cannot be improved in itself; it demands only to be listened to. But the poet can make himself better prepared for or available to it by removing obstacles and noise. He can guard and protect it, and thus foster the spontaneous progress of its strength and purity in him. He can educate himself to it by never betraying it.[17]

Many writers and artists have commented on the quality of knowing that emerges from patient absorption. Kafka, in his 'Reflections', says: 'You do not need to leave your room. Remain sitting at your table and listen. Do not even listen, simply wait. Do not even wait, be quite still and solitary. The world will freely offer itself to you to be unmasked, it has no choice, it will roll in ecstasy at your feet.'[18] T. S. Eliot in 'East Coker' enjoins us to 'be still, and wait without hope/for hope would be hope for the wrong thing'.[19] Martin Heidegger's *Discourse on Thinking* puts it very clearly.

> Normally when we wait we wait *for* something which interests us, or can provide us with what we want. When we wait in this human way, waiting involves our desires, goals and needs. But waiting need not be so definitely coloured by our nature. There is a sense in which we can wait without knowing for what we wait. We may wait, in this sense, without waiting for anything; for anything, that is, which could be grasped and expressed in subjective human terms. In this sense we simply wait, and waiting may come to have a reference beyond [ourselves].[20]

Rainer Maria Rilke, in his *Letters to a Young Poet*, has this advice for his self-appointed poetic apprentice:

> If you hold to Nature, to the simplicity that is in her, to the small detail that scarcely one man sees, which can so unexpectedly grow into something great and boundless; if you have this love for insignificant things and seek, simply as one who serves, to win the confidence of what seems to be poor: then everything will become easier for you, more coherent and somehow more conciliatory, not perhaps in the understanding, which lags wondering behind, but in your innermost consciousness, wakefulness and knowing.[21]

Poetic sensibility is available to everyone. It is not the special preserve of Poets with a capital P: people who deliberately create those forms of words called 'poems'. To be a Poet it is necessary to see 'poetically': necessary, but not sufficient. In addition, the Poet must be able to use language in such a way that the reader of the poem is invited not just into the Poet's world, but into the same mental mode, the same slow, poetic way of knowing, that gave rise to the poem in the first place. When we look at things in their own right, without referring them immediately to our own self-interest – which is what the poet invites us to do – then we are in a mode of sensing, knowing and learning that can reveal to us aspects of the world that lie outside the perimeter of our intentions and desires. In fact it can give us self-knowledge by situating our concerns within a wider context that is normally obscured. By allowing the poem to suck us in, we are drawn into a mode of perception that is situated upstream of our usual habits of conceptualization and self-reference. Simultaneously we know the world, and we know ourselves, differently. A poem, viewed thus, is a device for inducing a specific kind of sensibility in the reader. In Paul Valéry's terms, a poem is 'a kind of machine for producing the poetic state of mind by means of words'.

The Poet achieves her effect by doing two things at once. She paints a picture that invites our interest, our engagement, and our identification. And she does this with language that hampers our habitual ways of construing. We cannot see through our own system of categories and concerns without grossly violating the poet's words and thus we hang motionless for a moment in the presence of something made strange and new. George Whalley, writing about the 'teaching' of poetry in school, emphasises the vital necessity of 'experiencing' the poem, by which he means 'paying attention to it as though it were not primarily a mental abstraction, but as though it were designed to be grasped directly by the senses, inviting us to function in the perceptual mode'.[22] If we present a poem, especially to young minds, as something to be 'interpreted' and 'explained', as a kind of mental problem to be solved, like an extended crossword puzzle, we have missed the point. Reading poetry is an exercise in 'holding cognitive activity in the perceptual mode'. One must not *search* for meaning, but marinate oneself in the poem, so to speak, and let meaning come. If one does not treat the poem respectfully, as if it had a life and an integrity of its own, one ends up constructing a surrogate poem as a plausible substitute for the real one: one that disconcerts you less, and merely gives you back your own familiar code of conduct and comprehension.

A poem that is grasped intellectually generates a certain cerebral satisfaction. But a poem with which one is really engaged creates a bodily frisson of undisclosed import; a visceral and aesthetic response, and not just a mental one. Just as with the process of focusing, the body feels something that the mind may not understand. A. E. Housman illustrates the physicality of poetry, as we might expect, with power and humour:

> Poetry indeed seems to me more physical than intellectual. A year or two ago I received from America, in common with others, a request that I would define poetry. I replied that I could no more define poetry than a terrier could define a rat,

> but that I thought we both recognise the object by the symptoms which it provokes in us . . . Experience has taught me, when I am shaving of a morning, to keep watch over my thoughts, because, if a line of poetry strays into my memory, my skin bristles so that the razor ceases to act. This particular symptom is accompanied by a shiver down the spine. There is another which consists in a constriction of the throat, and a precipitation of water to the eyes. And there is a third which I can only describe by borrowing a phrase from one of Keat's last letters, where he says, speaking of Fanny Brawne, 'everything that reminds me of her goes through me like a spear.'[23]

Benedetto Croce, writing in the early years of this century, attempts in his *Aesthetic* to make the response in terms of beauty the linchpin of his approach to intuition.[24] For Croce, beauty is not a property of objects or of nature, but of an intuitive response. For the viewer of a painting or a sculpture or a dance, as much as for the reader of a poem, the aesthetic response is the felt manifestation of a certain way of seeing, or knowing, which that object has succeeded in inducing. That which is seen just as it is, fully attended to, not subsumed by categories or reduced to labels, is beautiful. One must learn to recognise, tolerate, enjoy and eventually value this intrinsic ambiguity and impenetrability; in Louis MacNeice's phrase, 'the drunkenness of things being various'.[25] Poetic sensibility and intuition are richer, fuller and subtler than everyday language. There are forms of knowledge that defy articulation. Impressions speak and resonate as vibrant wholes, undismembered. In this way of knowing, beauty, truth and ineffability come together. Argentinean writer Jorge Luis Borges, for example, adumbrating some of the natural 'attractors' of the poetic mode of mind, suggests that:

> Music, states of happiness, mythology, faces belaboured by time, certain twilights and certain places try to tell us something, or have said something we should have missed, or are about to say something: *this imminence of a revelation which does not occur is, perhaps, the aesthetic phenomenon.*[26] (Emphasis added)

Exquisite though the poetic way of knowing may be, we should not be seduced into desiring it as a permanent replacement for mundane reason. It remains one mental mode among many, and to be trapped in the poetic mode would be as disastrous as to be trapped in d-mode. Neurologist Oliver Sacks, in *The Man Who Mistook his Wife for a Hat*, recounts the moving story of one of his patients who was in just this position.[27] Rebecca, at nineteen, was unable to find her way around the block, could not confidently use a key to open a door and sometimes put her clothes on back to front. She had difficulty understanding straightforward sentences and instructions and could not perform the simplest calculations. Yet she loved stories and especially poetry, and seemed to have little difficulty following the metaphors and symbols of even quite complex poems. 'The language of feeling, of the concrete, of image and symbol formed a world she loved and, to a remarkable extent, could enter.' She performed appallingly on standard neurological tests, which are, as Sacks perceptively notes, specifically designed to deconstruct the whole person into a stack of 'abilities'. And just because of this, the tests gave no inkling of 'her ability to perceive the real world – the world of nature, and perhaps of the imagination – as a coherent, intelligible, poetic whole'. In the domain of conscious, deliberate intelligence she was severely handicapped. In the pre-conceptual, unreflective world, she was healthy, happy and competent.

At first, Sacks suggested she should attend classes to try to improve some of her basic 'skills', but they were of no use as they inevitably fragmented her. As Rebecca herself said: 'They do nothing for me. They do nothing to bring me together . . . I'm like a sort of living carpet. I need a pattern, a design, like you have on that carpet. I come apart, I unravel, unless there's a design.' And indeed, when she was spontaneously absorbed in an activity that engaged all of her, she was a different person. She was moved from the 'remedial classes' to a theatre workshop – which she loved and where she blossomed. She became *composed*, complete, and played her roles with poise, sensitivity and style. Sacks concludes his account: 'Now, if one sees Rebecca on stage, for theatre soon became her life, one would never even guess that she was mentally defective.' Lost in the particular, Rebecca became fluent and complete. In the world of the abstract, she was shattered and lost.

Skimping on perception runs two risks. Not only may one overlook aspects of the inner and outer world that are informative or even inspiring; one may inadvertently stir into perception as it develops assumptions and beliefs that are not justified or required. What is finally served up to consciousness may be simultaneously impoverished and elaborated, even adulterated. The mind in a hurry tends to see what it expects to see, or wants to see, or what it usually sees. One of the problems with the name 'Guy' is that I am forever reacting to calls that were not meant for me. People shouting 'Hi!' or 'Bye!' in crowded streets are likely to find me looking at them expectantly – before I detect and correct my perceptual mistake. Leaping to conclusions in this way is a gamble. By setting the threshold for the recognition of my name on a hair trigger, I make sure I react quickly – but I also make a lot of 'false positive' errors. By assuming that what usually happens is what did happen, I save processing time, but at the cost of misdiagnosing the situation when it is unusual. The fourth manner of paying attention which I want to describe in this chapter is a way of seeing through one's own perceptual assumptions. It is called mindfulness.

The extent to which the world-as-perceived is a mirror of our preconceptions and our preoccupations (and therefore the extent to which our subsequent thoughts, feelings and reactions are assimilated by these assumptions) is easy to underestimate. It takes an effort to see what is happening, because our beliefs are dissolved in the very organs which we use to sense. Take, as a trivial example, saliva. Be aware, for a moment, of the saliva in your mouth. Collect a little and roll it around. Feel how it lubricates your tongue as it slides over your teeth. Now get a clean glass, spit some of this saliva into it – and drink it. Notice how your perception of, and attitude towards, the same substance has miraculously changed. What was 'clean' and 'natural' has, through its brief excursion beyond the body, turned into something 'dirty' and 'distasteful'. The spit has not changed; only the interpretation.

One of the major contributions of experimental psychologists this century has been to keep providing us with new and telling demonstrations of what they call the 'theory-ladenness' of perception (just as it has been the function of the poets throughout history to keep showing us that the world is more 'various', more open to reinterpretation, and more inscrutable than we normally suppose). Much of the work on visual illusions shows this clearly. In the Kanizsa figures below, for example, we see – literally see – shapes that are not 'really there', because it seems plausible to the mind to suppose that they are.[28] We are used to seeing as 'whole', objects parts of which are occluded by other objects in front of them. And this expectation can drive us, if it 'makes sense' to do so, to hallucinate an intervening shape, even creating visible edges for it, adding impressions of depth and contrasts in brightness, to make the interpretation more convincing. Such tinkering with reality goes on

Figure 12. Illusory shapes and contours, after Kanizsa (1979)

all the time, and at levels of mind that are way below conscious intention or control.

A less stylised example is provided by the concept of 'old age'. Being 'old' is not just a biological phenomenon; how one goes about 'being old' depends on one's (largely unconscious) *image* of what it is like, what it means, to be old, and this in turn reflects a whole raft of both cultural assumptions and individual experiences. Ellen Langer and colleagues at Harvard University have examined the effect on elderly people of their own vicarious experiences, as children, of ways of being old. They reasoned that children may unconsciously pick up images of old age from their own grandparents – which they might then recapitulate as they themselves get older. Specifically, they surmised that the younger their grandparents were when children first got to know them, the more 'youthful' would be the image of old age that the children would unconsciously absorb, and the more positively they would therefore approach their own ageing.

In order to test this idea, they interviewed elderly residents of nursing homes in the vicinity of Boston to find out if they had lived with a grandparent as they were growing up and, if so, how old they were when the grandparent first moved in. When they were independently evaluated by nurses who knew nothing about the research, it was found that those elderly people who had lived with a grandparent when they themselves were toddlers were rated as more alert, more active and more independent than those whose first experience of living with a grandparent had not occurred till they were teenagers. While further research is needed to clarify the interpretation of these results, it does look as if the ways in which different people age depends quite directly on the assumptions and beliefs they have picked up in their own childhoods about what it is to be old.[29]

The unconscious assumptions that people stir into their experience are often rather hard to alter, but sometimes they can be changed just by a suggestion, especially if it comes from some kind of authority figure. The experience of pain, for instance, can be dramatically altered, in normal conscious subjects, simply by telling them to think of it differently. When a group of people who had volunteered to suffer some mild electric shocks were told to think of the shocks as 'new physiological sensations', they were less anxious, and had lower pulse rates, than those who were not so instructed.[30] In another study, hospital patients who were about to undergo major surgery were encouraged to realise how much the experience of pain depends on the way people interpret it. They were reminded, for example, that a bruise sustained during a football match, or a finger cut while preparing dinner for a large group of friends, would not hurt as much as similar injuries in less intense situations. And they were shown analogous ways of reinterpreting the experience of being in hospital so that it was less threatening. Patients who were given this training took fewer pain relievers and sedatives after their operations, and tended to be discharged sooner, than an equivalent group who were untrained.

These experiments demonstrate how other people may be able to rescue us from what Langer refers to as 'premature cognitive commitments' – help us become aware of the assumptions that we had dissolved in perception, and contemplate alternative ways of construing the situation. Helping others to change not the circumstances of their lives but their interpretations of those circumstances is a widespread therapeutic technique called 'reframing'. R. D. Laing, for example, in a classic case, helped a man who was desperate about his 'insomnia' to reconstrue his extra hours of wakefulness as a boon. 'Just think of all those people out there who are suffering from "somnia", forced to spend as much as eight or nine hours every day doing nothing,' Laing observed. When the 'problem' that we are facing is created by our own unconscious additives, no amount of good thinking or earnest effort will bring a solution. Such contortions only compound the original mistake. The only way out of the trap is to *see through* the interpretation which one had been making; to see it *as* an interpretation. Only with such self-awareness, or 'mindfulness', is it possible to be released from the pernicious belief.

Mindfulness involves observing one's own experience carefully enough to be able to spot any misconceptions that may inadvertently have crept in. There are a number of ways in which this quality of mindfulness towards the activity of our own minds can be cultivated, though all involve slowing down the onrush of mental activity, and trying to focus conscious awareness on the world of sensations, rather than jumping on the first interpretation that comes along and hurtling off in the direction of decision and action. Mindfulness can be taught directly, as a form of secular meditation, for example. Jon Kabat-Zinn, Director of the Stress Reduction Program at the University of Massachusetts Health Center, gives a clear idea of what is involved:

> The essence of the state is to 'be' fully in the present moment, without judging or evaluating it, without reflecting backwards on past memories, without looking forward to anticipate the future, as in anxious worry, and without attempting to 'problem-solve' or otherwise avoid any unpleasant aspects of the immediate situation. In this state one is highly aware and focused on the reality of the present moment 'as it is', accepting and acknowledging it in its full 'reality' without immediately engaging in discursive thought about it, without trying to work out how to change it, and without drifting off into a state of diffuse thinking focused on somewhere else or some other time . . . The mindful state is associated with a lack of elaborative processing involving thoughts that are essentially *about* the currently experienced, its implications, further meanings, or the need for related action. Rather mindfulness involves direct and immediate experience of the present situation.[31]

There is now good evidence for the efficacy of such mindfulness training in helping people with all kinds of distresses and diseases. Kabat-Zinn's programme has enabled hundreds of people with painful and upsetting conditions to release the secondary fears and anxieties that such conditions invariably create. Even the painfulness of pain itself, as we have just seen, can be reduced through mindfulness.

One particularly compelling demonstration of the practical value of mindfulness comes from clinical psychologist John Teasdale in Cambridge, who has been working on ways to prevent relapse in people who suffer from chronic depression.[32] To simplify a complex story: in many types of depression people suffer some upsetting experience or feeling, but instead of just dealing with it as best they can, and moving on, a set of negative assumptions is activated which then triggers a downward spiral of pessimistic thoughts, memories, feelings and interpretations. Once this process has taken hold, people come to see the world and themselves through increasingly critical glasses, and this makes it all the more likely that they will attend to just those features of their experience that validate and exacerbate their feelings of inadequacy or hopelessness. It becomes impossible to remember, or even to notice, anything positive or encouraging. The conscious mind may become obsessed with 'personal goals that can neither be attained nor relinquished'.

Teasdale argues that the way to stop this vicious spiral from getting going is not to try to prevent experiences of disappointment or uncertainty: that is not a practical option. There will always be upsets. Rather the solution is to get people to practise new habits of thinking and attending which will stop the self-destructive patterns from gaining control of the mind. And mindfulness can do this by preventing you from leaping to conclusions, and then carrying on as if those conclusions were solid and true: first of all by keeping you closer to the 'bare facts' and enabling you to see molehills *as* molehills, rather than automatically inflating them into mountains; and secondly, as you become more attentive to the movements of the mind, you relearn your attitude towards them. The conclusions that present themselves to consciousness are not seen any more as 'valid descriptions of who I really am' – 'worthless', for example – but as 'thoughts produced by the mind'. You reinterpret them as 'mental states', or 'events in the field of awareness', not as 'reflections of reality'. So even when negative interpretations and conclusions do bubble up, mindfulness enables you to refuse the lure and question their validity. It is no longer 'me' who is forced to defend myself; it is the content of consciousness that now appears dubious. The tables are turned.

John Teasdale's conclusions from his research with depressive people may 'ring bells', perhaps, with a wider population.

> Depressive relapse often seems to occur when patients fail to take appropriate remedial or coping activity at an early stage of incipient relapse, when control over depression is likely to be relatively easy to obtain. Patients may defer recognition or

> acknowledgement of problems to a later stage in the relapse process, where a more full-blown depressive syndrome may be much more difficult to deal with . . . Mindfulness training . . . in 'turning towards' potential difficulties, rather than 'looking away' from them, is likely to facilitate early detection of signs . . . and so to increase the chances that remedial actions will be implemented at a time when they are likely to prove most effective.

Though pharmacological approaches to depression continue to play a vital role in its amelioration, the research shows that Teasdale's and Kabat-Zinn's approach is at least as effective as administering conventional anti-depressant drugs (and there are, of course, fewer negative side effects).

Daniel Goleman, in *Emotional Intelligence*, has documented the role that mindfulness can play in preventing 'emotional hi-jacking'.[33] When couples begin to get into a marital 'fight', for example, things can easily go from bad to worse if either or both of the partners falls into a self-reinforcing pattern of negative thinking. Mindfulness increases the likelihood that such a pattern can be spotted and neutralised before it has done too much damage. Goleman gives the example of a wife who feels in the heat of the moment that 'he doesn't care about me or what I want; he's always so selfish', but who, on catching herself in the act of 'demonising' her husband, is able to remind herself that 'There are plenty of times when he has been caring – even though what he did just now was thoughtless and upsetting.' Through the moment of mindfulness she is able to neutralise the exaggerated thought that, if accepted, would have justified a negative reaction that would only have inflamed the situation further.

The value of slowing down the mind is evident in dozens of everyday situations. Take the example of a divorcing couple arguing over custody of their child. In such an emotionally intense situation, it is very easy for an impoverished perception to lead to a rigid response: one in which it looks as if only one partner can 'win', and the other therefore must 'lose'. Any more subtle analysis of the actual predicament, and especially of what it is that each party is actually trying to achieve, is sacrificed in favour of a knee-jerk adherence to a one-dimensional view. Underneath the dogmatism, however, there may be a whole host of other factors and values, consideration of which might make it possible for everybody – including the child – to win. Is each parent really seeking the full-time physical presence of the child, or is it a quality of relationship they want to preserve? Are they trying to use the custody issue as a way of punishing their partner, or of asserting a need for control that they feel they lost in the marriage? Could there not be advantages to being a part-time parent that have been overlooked? And what might be best for the child itself? Increasingly it is the role of counsellors and mediators to try to ease people out of their entrenched, antagonistic positions, and to see the situation more fully. To cultivate mindfulness is to be able to adopt that role for oneself.

The cultivation of mindfulness does not require instruction in any kind of formal meditation, though it may help. Our own culture possesses many venerable and effective activities – or inactivities – that encourage the mind to shift out of doing-and-thinking mode into a mode that is relaxed and spacious, yet alert to its own meanderings. Coarse fishing, for example, as Ted Hughes noted, is a meditation in everything but name; a perfect excuse to gaze at the float while the mind wanders free, enjoying the shimmer of the light on the water or the soft touch of the rain. It is not unknown for fishermen to experience mild resentment when their reverie is disturbed by the inconvenient attentions of fish to hook. Rhythmic activities such as knitting, weeding and swimming may all encourage mindfulness of simple body sensations, sounds or smells, drawing attention away from problem-solving and back into the perceptual world. Watching a relatively unimportant county cricket game live is good practice. On TV you can drift off too much, knowing that the action replays will show you the highlights. At the ground, if you don't maintain awareness you miss the action when it occasionally happens. Yet you cannot spend all day focused and concentrated. You have, gradually, to develop the quality of attention of a cat: relaxed and watchful at the same time. You feel the spontaneous 'pulsing' of awareness that we spoke of in Chapter 10.

If perception samples experience only in order to categorise it, and to decide whether it is potentially useful or harmful, the conscious image it creates is likely to be rather flat and dull. Having lost perceptual vividness, we seek to put ourselves in extreme situations where the outside world startles us strongly enough for perceptual intensity to return. Thus the entertainment 'industry', whether in the form of violent or pornographic films, terrifying theme-park rides, raves or cocaine, becomes geared to providing transient experiences of aliveness that our habitual mode of mind prevents us from having for, and by, ourselves. Greater mindfulness makes conscious experience of life richer and more vivid. Rediscovering the ability to dwell in perception gives it back its charm and its vitality.

CHAPTER 12

The Rudiments of Wisdom

The wisdom of a learned man cometh by opportunity of leisure; and he that hath little business shall become wise.

Ecclesiasticus 38:34

Summerhill, the progressive English school founded by A. S. Neill, is run by a council, the 'moot', which meets once a week. Every member of the school, from the newest five-year-old to the oldest teacher, has a single, equal vote. The moot decides everything: school rules, bedtimes, sanctions to be applied in particular cases. It is the mid-1970s. The founder and his wife sit quietly in the meeting listening to a complaint being brought by two girls against one of the boys who has, allegedly, been irritating them by, among other things, flicking them with towels. The mood of the meeting is against the defendant. Student after student denounces him. A harsh penalty looks likely. Both Neill and his wife sit with their hands raised, waiting to be invited to speak by the thirteen-year-old girl who is chairing the moot. Eventually Mrs Neill has her turn. 'Just think how dull your lives would be if you didn't have these boys to annoy you,' she says, with a twinkle. The meeting laughs. Then Neill speaks in a gruff, laconic voice, as if he is raising a point of procedure. 'I don't think the meeting has any right to interfere in a love affair,' he says. Again the meeting laughs. The boy and one of the girls grin sheepishly at each other. The meeting moves on.[1]

The exploration of 'knowing better by thinking slower' eventually brings us to a consideration of wisdom. The dictionary tells us that wisdom is 'the capacity to judge rightly in matters relating to life and conduct; sound judgement especially in practical affairs; making good use of knowledge'. But that does not get us very far. What does it mean to 'judge rightly', or to have 'sound judgement'? Who is to decide what is right or sound? What sort of knowledge does one need, and how does one learn to make good use of it? All the interesting questions are begged. Our study of the complex and sometimes troubled relationship between the hare brain and the tortoise mind can help us to get a better handle on this most elusive, but most important, of concepts.

The Neills' reactions demonstrate some of the qualities of wisdom. Above all, wisdom is practical, dealing directly with 'matters relating to life and conduct'; with 'practical affairs'. It is also creative and integrative. The Neills 'reframe' a polarised situation in a way that skilfully avoids taking sides. Where the protagonists are stuck in a world-view in which one must 'lose' if the other 'wins', the wise counsellor finds a perspective that integrates and transcends the opposing positions. Apparently stark choices are magically transformed into common purposes. A classic example of this creative reframing occurred during one of the many riots in nineteenth-century Paris, when the commander of an army detachment was ordered to clear a city square by firing at the *canaille* – the rabble. He commanded his soldiers to take up their firing positions, their rifles levelled at the crowd, and, as a ghastly silence descended, he drew his sword and shouted at the top of his lungs: 'Mesdames et messieurs, I have orders to fire at the *canaille*. But as all I can see from here are a great number of honest, respectable citizens peacefully going about their lawful business, may I request that they clear the square quietly so that I can safely pick out and shoot the wretched *canaille*.' The square was emptied in a few minutes, with no loss either of life or face.[2]

Wisdom often involves seeing through the apparent issue to the real issue that underlies it. Where the Summerhill students saw only conflict, Neill saw a much more complex dynamic that included affection and playfulness in addition to the superficial disgruntlement; while his wife gently hinted at a longer-term perspective within which enacting and resolving such minor conflicts constituted an important and proper part of the 'curriculum' of growing up. The girls are reacting from a level of irritation which is genuine but also incomplete, less than the whole story, and Mrs Neill gently reminds them of a larger set of values which they share, but have temporarily forgotten. There is a part of the girls that would indeed be disappointed if the boys were to leave them in peace. On a grander scale, Nelson Mandela, in his famous inaugural speech as president of South Africa in 1994, sought to reframe the fears, and the aspirations, of his countrymen and women. He attempted to peel away one layer of understanding to reveal another.

> Our deepest fear is not that we are inadequate. Our deepest fear is that we are powerful beyond measure. It is our light, not our darkness, that most frightens us. We ask ourselves, who am I to be brilliant, gorgeous, talented and fabulous? Actually, who are you not to be? . . . There's nothing enlightened about shrinking so that other people won't feel insecure around you . . . As we let our own light shine, we unconsciously give other people permission to do the same. As we are liberated from our own fear, our presence automatically liberates others.

We might say that wise people are able to act and judge 'rightly' because they see through the complicated intermediate layers of value in which people sometimes become enmeshed to the simple truths and concerns that animate almost everyone: to feel safe; to express oneself without fear; to understand one's place and purpose in the world; to act with integrity; to belong somewhere; to love and be loved. As the French psychologist Gisela Labouvie-Vief has concluded from her study of wisdom, 'What makes the artist, the poet or the scientist wise is not expert technical knowledge in their respective domains but rather knowledge of issues that are part of the human condition. Wisdom consists, so to say, in one's ability to see through and beyond individual uniqueness and specialisation into those structures that relate us to our common humanity.'[3]

Wise judgements take into account not just ethical depth but the social and historical repercussions that may ensue. An expedient solution may follow from a partial analysis of a problem that represents only one point of view, or excludes a long-term perspective. Those attempting to conduct ethical business, for example, try to make decisions that benefit a constituency of 'stakeholders' that includes employees and their families, customers and local residents, as well as the shareholders, and which respect the interests and rights of future generations. Wisdom works with 'the big picture', one that accurately incorporates the moral, practical and interpersonal detail, however inconvenient, and tries to find a solution that fits and respects this complexity as well as possible. Wisdom does not search the rule-book for templates and generalities that the situation can be forced to fit. It tends to go back to the moral and human basics and custom-build a response that reconciles as many of the constraints and desiderata as possible.

Wisdom is uncompromising about fundamental values, but flexible and creative about the means whereby they are to be preserved or pursued – sometimes surprisingly or even shockingly so. A Zen master astounded his monks by burning statues of the Buddha to keep warm. Jesus cut through a convoluted moral predicament by telling his confused and angry followers to obey the law, but keep their spirits free, by 'rendering unto Caesar the things that are Caesar's, and unto God the things that are God's'. A wise action may seem to disregard convention, or even rationality. In desperate situations, where all other avenues are blocked, it may be wise to do something apparently absurd.

When in 1334 the Duchess of Tyrol encircled the castle of Hochosterwitz she knew that the fortress, built on a steep rock rising high above the valley floor, was impossible to attack directly, and would yield only to a long siege. And so it proved. Eventually both the defenders and the Duchess's troops were on the verge of giving up. The defenders were down to their last ox and their last two bags of barley. The attacking soldiers were becoming bored and unruly, and there was pressing military business elsewhere. The commander of the castle, at this point, seemed to lose his sanity. He ordered the ox to be slaughtered, its carcass to be stuffed with the barley, and the body thrown over the ramparts, whence it rolled down the cliff and came to rest in front of the enemy camp. Upon receiving this disdainful message, and assuming that anyone who could afford such an extravagant gesture must be well provisioned and in good heart, the discouraged Duchess gave up the siege and moved on.

If a predicament can be solved by d-mode, it does not need wisdom. Wisdom has been defined as 'good judgement in hard cases'.[4] Hard cases are complex and ambiguous; situations in which conventional or egocentric thinking only results in heightened polarisation, antagonism and impasse. In hard cases personal values may conflict: to choose the course of honesty is to risk the sacrifice of popularity; to choose adventure is to jeopardise security. As with the 'bystander studies' which I mentioned earlier, publicly to go to someone's aid may risk your being late for your appointment, getting your clothes dirty, or looking a fool when the situation turns out to be a student prank. Hard cases are those where important decisions have to be made on the basis of insufficient data; where what is relevant and what is irrelevant are not clearly demarcated; where meanings and interpretation of actions and motives are unclear and conjectural; where small details may contain vital clues; where the costs and benefits, the long-term consequences, may be difficult to discern; where many variables interact in intricate ways.

The conditions in which wisdom is needed, in other words, are precisely those in which the slow ways of knowing come into their own. To be wise is to possess a broad and well-developed repertoire of ways of knowing, and to be able to deploy them appropriately. To be able to think clearly and logically is a constituent of wisdom, but it is not enough on its own; many unwise decisions have been made by clever people. One needs to be able to soak up experience of complex domains – such as human relationships – through one's pores, and to extract the subtle, contingent patterns that are latent within it. And to do that, one needs to be able to attend to a whole range of situations patiently and without comprehension; to resist the temptation to foreclose on what that experience may have to teach. (The poet and critic Matthew Arnold, during his time as an inspector of schools, used to tell of a colleague who boasted of thirteen years' experience – whereas, Arnold would comment, it was perfectly clear to anyone who knew the man that he had had nothing of the sort. He had had one year's experience thirteen times.) And one must be able to take one's time: to mull over a problem and to dwell on details and possibilities. In short, to be wise one needs the tortoise as well as – perhaps even more than – the hare.

Allowing oneself time to be wise is vital in the context of caring professions such as counselling and psychotherapy. Robin Skynner, co-founder of both the Institute of Family Therapy and the Institute of Group Analysis, and author, with comedian John Cleese, of the books *Families and How to Survive Them* and *Life and How to Survive It*, has talked of his perennial confusion on working with a new group or a new family.[5] Even with more than forty years' experience, it regularly happens, he says, that a few minutes into the first consultation he feels lost. Suddenly his accumulated knowledge and skill appear to desert him. It seems as if he has no precedents on which to draw. He may wonder what he is doing there, or may even feel fraudulent. Nothing wise occurs to him to say or do. Yet, Skynner says, one of the major benefits of his vast experience is the courage not to flee from this barren state. It remains, after all this time, uncomfortable, yet he now recognises it to be an essential 'winter' phase, in which nothing seems to be growing, which precedes the arrival of spring. After half an hour or so, some tentative inklings and intuitions begin to form, and gradually a new sense of being able to work with this unprecedented situation emerges. Skynner's knowledge of interpersonal dynamics does not manifest itself as fast and certain prescriptions: far from it. It appears through the courage to wait, and to notice and trust the fragile shoots of understanding that eventually start to appear.

The way of knowing that generates wisdom is a curious one, for it seems to transcend conventional dualities. It is at once subjective and objective; both involved, caring and affectionate, and yet dispassionate and unclouded by personal sentiment or judgement. There is what I have called 'poetic sensibility', in which the object of attention is known intimately, even 'lovingly', but without projection: no hopes or fears intrude to obscure the clarity of perception. If the 'object' is another person, someone in distress, for example, the wise counsellor is touched by their predicament, and yet untouched by it. She feels the situation as a human, and not just as a technical, one; but her empathy does not dissolve into mere sympathy or, worse, collusion. She is able to see – with her mindfulness – beliefs and opinions, both her own and others', *as* interpretations, and not – as they may appear to the sufferer – as the transparent truth.

Psychotherapist Carl Rogers described empathy as:

> entering the private perceptual world of the other and becoming thoroughly at home in it . . . It means temporarily living in his/her life, moving about in it delicately without making judgments . . . as you look with fresh and unfrightened eyes at elements of which the individual is fearful . . . To be with another in this way means that *for the time being* you lay aside the views and values you hold for yourself in order to enter another's world without prejudice.[6]

To perform this delicate balancing act, the wise person needs to be mindful not only of the other's world, but of his own as well. As in focusing, he needs to be able to 'tune in' to his own inner state to ensure that no judgements or projections are slipping unnoticed into his own interpretation of the situation. Only if his perception is clean and full will his judgement be subtle, fair and trustworthy. That is why the so-called 'counter-transference' in psychotherapy – projections of affection or even sexual attraction, for example, on to the client by the therapist – is such an important issue (and why doctors, in the United Kingdom at least, are not permitted to treat members of their own families). But this dispassionate yet kindly vantage point is not easily achieved. As the Danish philosopher Kierkegaard said: 'The majority of men are subjective towards themselves and objective towards all others – terribly objective sometimes

– but the real task is in fact to be objective towards one's self and subjective towards all others.'[7]

There is little empirical research on wisdom itself; but there is some information about what people consider wisdom to be. Yale psychologist Robert Sternberg summarised the general view of the wise individual thus:

> [She or he] listens to others, knows how to weigh advice, and can deal with a variety of different kinds of people. In seeking as much information as possible for decision-making, the wise individual reads between the lines . . . The wise individual is especially able to make clear . . . and fair judgements, and in doing so, takes a long-term as well as a short-term view of the consequences of the judgement made . . . [She or he] is not afraid to change his or her mind as experience dictates, and the solutions that are offered to complex problems tend to be the right ones.[8]

The ability which mindfulness brings, and which wisdom seems to presuppose – to see one's own knowledge, as well as that of others, as a personal and social construction, capable of being interrogated, reframed or reconstrued – is not easily developed, nor does it come without cost. It requires a considerable sense of personal security to give up the belief in certain knowledge. It is not just the admission that one's knowledge is always incomplete – that there is always more that one could consider – which is required, but the recognition that knowledge itself is essentially unsure, equivocal, open to question and reinterpretation. As Harvard educationalist Robert Kegan has recently reminded us in his book *In over our Heads: the Mental Demands of Modern Life*, this perspective is only gained at the cost of 'a human wrenching of the self from its cultural surround'.[9] Adult educators, for example, who are demanding this reflective, critical ability from their students, are asking

> them to change the whole way they understand themselves, their world, and the relation between the two. They are asking many of them to put at risk the loyalties and devotions that have made up the very foundation of their lives. We acquire 'personal authority', after all, only by relativizing – that is, only by fundamentally altering – our relationship to public authority. This is a long, often painful voyage, and one that, for much of the time, may feel more like mutiny than a merely exhilarating expedition to discover new lands.

To be wise requires the development of a mode of mind which can accept the relative nature of knowledge without tipping into rampant subjectivity or solipsism. One must be able to live with Voltaire's dictum: 'doubt is an uncomfortable condition, but certainty is a ridiculous one'. Yet this doubt must leave freedom to act – sometimes quickly and decisively. The wise person walks a narrow line between the twin perils of rigid dogmatism and paralysing indecision. As psychologist John Meacham has put it: 'one abandons both the hope for absolute truth and the prospect that nothing can be known; in wisdom, one is able to act with knowledge while simultaneously doubting.'[10] Meacham makes his point with an example in which this awareness of fallibility is conspicuously absent. In the film *The Graduate*, the young man Ben (Dustin Hoffman) is taken aside at his graduation party by his father's friend Mr Maguire. 'Come with me for a minute,' says Maguire, 'I want to talk to you. I just want to say one word to you. Just one word.' 'Yes, sir,' Ben replies. 'Are you listening?' insists Maguire. 'Yes, sir, I am,' says Ben. 'Plastics,' says Maguire. There is a long pause while they look at each other. Finally Ben asks, 'Exactly how do you mean that, sir?' The humour of this laconic scene turns precisely on Maguire's unwise conviction that his knowledge, so generously offered, is absolutely incorrigible. Meacham points out that an epistemological milieu, whether it be school, university or workplace, that requires one to appear certain militates against the development of wisdom. On the contrary, 'An intellectual climate hostile to ambiguity and contradiction is one that encourages easy solutions such as stereotyping and intolerance.'

Wisdom arises from a friendly and intimate relationship with the undermind. One must be willing, like Winnie the Pooh, to 'allow things to come to you', rather than, like Rabbit, 'always going and fetching them'. D-mode clings to lines of thought that are clear, controlled, conventional and secure: precisely those to which 'hard cases', by their very nature, will not succumb. Wisdom comes to those who are willing to expand their sense of themselves beyond the sphere of conscious control to include another centre of cognition to which consciousness has no access, and over which there seems to be little intentional jurisdiction. As Emerson puts it: 'A man finds out that there is somewhat in him that knows more than he does. Then he comes presently to the curious question, Who's who? Which of these two is really me? The one that knows more or the one that knows less; the little fellow or the big fellow?'

Those sages and seers who represent the most clear-cut embodiments of wisdom tend to give two different answers to Emerson's question. Those that belong to the theistic religious traditions are inclined to retain their personal identification with the 'little fellow', and to assume that the 'big fellow' is some external source of authority who, through its grace and mercy, has chosen to speak 'through' them. In the Judeo-Christian tradition, for instance, the broadcasting authority is referred to as God. Yet even within these religions there have been dissenting voices which have insisted on seeing the source of wisdom as immanent. The 'big fellow' is still called God, or the Godhead, but is now construed as an inscrutable force or process that is located within. In the so-called 'apophatic' tradition within Christianity, for example, there are many mystics and sages who have expressed their rediscovered intimacy with the undermind in these terms. The founder of the apophatic tradition, the sixth-century Dionysius the Areopagite, described the mystic as one who 'remains entirely in the impalpable and the invisible; having renounced all knowledge [he] is united to the Unknowable – to God – in a better way, and *knowing nothing, knows with a knowledge surpassing the intellect*'.[11] To Dionysius, 'The most godly knowledge of God is that which is known by unknowing.'

Eckhart von Hochheim, 'Meister Eckhart' to history and to his followers, is acknowledged to be perhaps the greatest of the Christian mystics, though in his time his works were condemned by a papal commission as heretical and dangerous. He died only just in time to avoid being burnt at the stake, it appears. For Eckhart, 'A really perfect person will be so *dead to self*, so lost in God, so given over to the will of God, that his whole happiness consists in being *unconscious of self and its concerns*, and being conscious instead of God.' The goal of spiritual practice is to find the inner place 'where never was seen difference, neither Father, Son nor Holy Ghost, where there is no one at home, yet where the spark of the soul is more at peace than within itself'.

Johannes Tauler, a Dominican monk and disciple of Meister Eckhart who lived and taught in the German Rhineland in the mid-fourteenth century, was one of the first of the apophatics to adopt an explicitly psychological interpretation of his religious experience. His practical instruction on the contemplative life, delivered mostly in his sermons at the convents and monasteries which he visited, offers very clear advice to the nuns and monks on how to pursue their devotions inwardly. For him it was self-evident that human beings yearned for inwardness; for 'personal renewal through a submersion in the divine ground from which all creatures have arisen'.[12] And this divine ground was none other than the Kingdom of God.

> This Kingdom is seated properly in the innermost recesses of the spirit. When the powers of the senses and the powers of the reason are gathered up into the very centre of the man's being – *the unseen depths of his spirit, wherein lies the image of God* – and thus he flings himself into the divine abyss . . . where [everything] is so *still, full of mystery and empty*. There is nothing there but the pure Godhead. Nothing alien, no creature, no image; no form ever penetrated there.

Tauler's method for approaching this Kingdom relied on the deliberate, methodical cultivation of passivity, a way of turning oneself over to forces and impulses that did not originate from the sphere of the conscious self, which he, following Eckhart, and anticipating Heidegger, referred to as *Gelassenheit*, 'letting be'. Of the pragmatic benefits of this attitude Tauler was in no doubt: 'In this way nature and reason become purified, the head strengthened, and the individual more peaceful, more kind and more restful.'

Tauler's psychological interpretation of the Church's symbolism must also have seemed, to his more conventional contemporaries, to border on the blasphemous. In his view of the Trinity, for instance, no longer is God the Father a transcendent figure. He becomes the Godhead, the innermost source, the unconscious mystery. The Son represents the perpetual birth of 'something' out of this divine 'nothing', the amazing coming into being of conscious experience and physical acts, continually gushing forth from the impenetrable fountainhead. And the Spirit is the transformation of being, the wisdom, the 'peace of God which passeth all understanding', which is available to all those who are willing, as Eckhart put it, to 'naught themselves', to put their faith in the inner God who 'moves in a mysterious way'.

There is a surprising coming together of the direct insight of these sages (and one could quote from dozens more in similar vein) with the new science of the intelligent – even wise – unconscious. Perhaps it would encourage people to see that their own capacity for wisdom is amenable to cultivation if this confluence of understanding were to be more widely known. As Lancelot Law Whyte, one of the first to trace the history of the unconscious back into pre-Freudian times, concludes:

> Today faith, if it bears any relation to the natural world, implies faith in the unconscious. If there is a God, he must speak there; if there is a healing power, it must operate there . . . The conscious mind will enjoy no peace until it can rejoice in a fuller understanding of its own unconscious sources.[13]

Of all the major world religions, though, it is Buddhism that most clearly and consistently identifies the source of wisdom with the undermind. Indeed, Buddhism goes so far as to say that wisdom resides in the recognition that *all* the activities and contents of consciousness are merely manifestations of unconscious processes. Even our most rational and well-considered trains of thought are not created by the conscious 'I', but are merely displayed in consciousness, like images and text on a computer screen. The screen has no intelligence of its own; it merely portrays the results of a certain kind of activity within the unobservable world of the microchips. The Buddhist project, we might say, is to bring about a shift in our identity's 'centre of gravity' from consciousness to the mysterious undermind. Through the meticulous form of attention that is cultivated by mindful meditation, we become more fully aware of the passing details of experience, and the wayward, ephemeral and suspect nature of the conscious mind begins to become more evident. One has to be able to adopt a more sceptical attitude towards consciousness itself, reconceptualising it as a drama, a passing show, rather than as a reliable star to steer by. In the words of a contemporary Tibetan *dzogchen* master:

> Whatever momentarily arises in the body-mind . . .
> has little reality.
> Why identify and become attached to it,
> passing judgement on it, and ourselves?
> Far better to simply
> let the entire game happen on its own,
> springing up and falling back like waves.[14]

The relocation of the centre of both identity and intelligence to the undermind is expressed, within Buddhism, most clearly in the Zen tradition. The contemporary Japanese teacher Shunryu Suzuki Roshi, who founded and for many years directed the Zen Centre in San Francisco, used to say: 'In Japan we have the phrase *shoshin*, which means "beginner's mind". The goal of practice is always to keep our beginner's mind . . . If your mind is empty, it is always ready for anything; it is open to everything. In the beginner's mind there are many possibilities; in the expert's mind there are few.'[15] Another contemporary Zen teacher, the Korean Seung Sahn Sunim, who has also made his home in the United States, instructs his students thus:

> I ask you: What are you? You don't know; there is only 'I don't know'. Always keep this don't-know mind. When this don't-know mind becomes clear, then you will understand. So if you keep don't-know mind when you are driving, this is driving Zen. If you keep it when you are talking, this is talking Zen. If you keep it when you are watching television, this is television Zen. You must keep don't-know mind always and everywhere. This is the true practice of Zen.[16]

But, as in Christianity, the discovery of the value of the hidden layers of the mind is not a modern achievement. As long ago as the seventh century, Chinese teacher Hui-Neng, the Sixth Patriarch of Zen, presumed author of the influential *Platform Sutra*, was encouraging his followers to take note of the activities of the undermind.[17] A bluff, down-to-earth fellow, by all accounts, he devoted much of his energy to trying to combat the prevalent idea that spiritual realisation involved cutting off your thoughts, and that if you meditated earnestly and frequently enough, 'enlightenment' would be your inevitable reward. Spiritual practice, for Hui-Neng, was not about calming or emptying the mind; it was about noticing, in any and every moment, in whatever you were up to, the dynamic relationship between conscious and unconscious.

> O friends, if there are among you some who are still in the stage of learners, let them turn their illumination upon the source of consciousness whenever thoughts are awakened in their minds . . . The [conscious] mind has nothing to do with thinking, because *its fundamental source is empty*. . . It is called 'ultimate enlightenment' when one has awakened to the source of the mind.

Buddhist scholar D. T. Suzuki explains very clearly what this awareness of the interface between conscious and unconscious means, in terms of the key Buddhist concept of *prajna*, usually translated as 'wisdom'.

> Prajna points in two directions, to the Unconscious, and to a world of consciousness which is now unfolded . . . When we are so deeply involved in the outgoing direction of consciousness and discrimination as to forget the other direction of

> Prajna, [wisdom] is hidden, and the pure undefiled surface of the Unconscious is now dimmed . . . Ordinarily the apperceiving mind is occupied too much with the outgoing attention, and forgets that at its back there is the unfathomable abyss of the Unconscious. When its attention is directed outwardly, it clings to the idea of an ego-substance. It is when it turns its attention within that it realises the Unconscious.

The undermind for Hui-Neng, like the Godhead for Tauler, is the 'nothing' that constantly brings forth the 'somethings' of the mind. The miracle, as D. T. Suzuki puts it, is that 'It is in the nature of Suchness [the Unconscious] to become conscious of itself . . . In the self-nature of Suchness there arises consciousness . . . Psychologically we can call [Suchness] the Unconscious, in the sense that all our conscious thoughts and feelings grow out of it.' And: 'To see self-nature means to wake up in the Unconscious.'

> When Ma-tsu and other Zen leaders declare that 'this mind is the Buddha himself', it does not mean that there is a kind of soul lying hidden in the depths of consciousness; but that a *state of consciousness . . . which accompanies every conscious and unconscious act of mind is what constitutes Buddhahood.*

CHAPTER 13

The Undermind Society: Putting the Tortoise to Work

The great spectre that recurrently haunted many of the most sensitive men of the last two hundred years is that there may eventually come a time when all the richness and amplitude of Creation will simply pass through the eyes of a man into his head and there be turned by the brain into some sort of formula or equation.

Nathan Scott

In the ballroom of the Washington Hilton, 1,500 top educators from around the world sit rapt while Doug Ross describes the future. Ross, then Assistant Secretary of Labor for Employment and Training in the Clinton administration, is delivering the closing address to the 1994 international conference 'A Global Conversation about Learning'. He is talking about the practical steps that the multi-billion-dollar agency he oversees is taking to bring about the 'learning society'. Plans already exist, he tells the audience, to create a system of tax incentives to encourage people to become lifelong learners: to invest in developing their own 'cognitive capital' throughout their working lives. Taking out a loan for learning should be as easy, and as attractive financially, as taking out a mortgage. New 'school to work' programmes are being designed to break down the traditional divide between academic and vocational study, and to bring sophisticated theory to life in the workplace.

Ross makes no bones about the fact that the new 'poor', the new marginals in society, will be those who cannot or will not learn. Learning and earning are already inextricably entwined, and becoming more so by the day. The work of the future will be overwhelmingly *mind* work. Manual and blue-collar workers, Ross claims, will constitute just 10 per cent of the workforce by early in the next century. Economic and vocational forces are placing unprecedented pressure on individuals throughout society continually to learn new knowledge and skills – and to develop their *ability* to learn: their confidence and their resourcefulness as learners, and their skill at managing their own learning lives.

The concern with learning is not just American: it is worldwide. In Britain a massive 'Campaign for Learning', supported by the government and big business, and coordinated by the Royal Society of Arts, was launched in 1996 to encourage people to, in the words of its chairman Sir Christopher Ball, 'care about their learning in the same way that we are all gradually learning to care about the environment or about our own personal health'. The aim of the campaign is 'to help create a learning society in the UK in which every individual participates in learning, both formal and informal, throughout their lives. This means boosting people's desire to learn, highlighting existing ways of learning and proposing new ones'. The campaign arose as a response to a large-scale survey which revealed that, while more than 80 per cent of the population believed that learning was important for them, less than a third had plans to do anything about it. For only £2.50 you can receive from campaign headquarters a pack of material that will enable you to create your own Personal Learning Action Plan, and many other initiatives are in the pipeline.

The pressure on individuals to become learners is not just a consequence of changes in the job market, however. If employers and governments can no longer offer everyone 'jobs for life', many other cultural sources of stability and authority have also been weakened and undermined over the course of the twentieth century. It is already something of a cliché to talk of 'the collapse of certainty', whether it be in terms of the disappearance of traditional communities, the rise of geographical mobility, the explosion in information and communication technologies, the interpenetration of different ethnic cultures, the appearance of a bewildering variety of new spiritual movements and leaders to challenge the authority of the orthodox religions, or the freedom to develop personal preferences and lifestyles that may bear no relationship to those into which one was born. For many people in Western society it is now not only possible to choose, to a considerable extent, how, where and with whom they are going to live, and who they are going to be: it is incumbent upon them to do so. Whether the freedom to invent oneself is experienced as welcome or unwelcome, the onus is on individuals to learn and grow, in all kinds of ways, as never before.

In the midst of these uncertainties and opportunities, it is, therefore, of paramount importance that people possess not just the confidence but the know-how to be able to learn *well.* Governments can create incentives and campaigns can exhort, but if people feel unsafe or unsupported, or are unskilled in the craft of learning itself, they may shy away from the learning opportunities that they encounter – even when such learning would clearly help them to pursue their own valued goals and interests. Learning, whether it involves mastering a new technology or recovering from a divorce, is a risky business, and a lack of either the tools or the self-assurance to pursue it results in stagnation.

In this context, it is all the more significant that cognitive science is currently drawing our attention to the curious fact that we have forgotten how our minds work. As we have seen, the modern mind has a distorted image of itself that leads it to neglect some of its own most valuable learning capacities. We now know that the brain is built to linger as well as to rush, and that slow knowing sometimes leads to better answers. We know that knowledge makes itself known through sensations, images, feelings and inklings, as well as through clear, conscious thoughts. Experiments tell us that just interacting with complex situations without trying to figure them out can deliver a quality of understanding that defies reason and articulation. Other studies have shown that confusion may be a vital precursor to the discovery of a good idea. To be able to meet the uncertain challenges of the contemporary world, we need to heed the message of this research, and to expand our repertoire of ways of learning and knowing to reclaim the full gamut of cognitive possibilities.

This will not be easy, for the grip of d-mode on late twentieth-century culture is strong, and reflects a trend in European psychology that has its origins right back in classical Greece. For Homer, the seat of human identity and intelligence was emotional rather than rational, and opaque rather than transparent. The undermind, *psyche*, was a living reality, experienced in the body, and interpreted either, as Julian Jaynes has suggested, as 'the voice of the gods', or as a vital ingredient of the human personality.[1] But by the time of Plato the centre of gravity of a person's being had drifted upwards into the head, and begun to become associated with reason and control. The undermind still existed as an emotional or intuitive force, but it had come to be seen as secondary and subversive, something wayward, primitive and unreliable. It was in reason that people were most truly, most nobly, themselves.[2]

This ambivalent relationship with the undermind was to continue for more than a millennium. Conscious reason might be the apotheosis of the self, but the acknowledgement of its more mysterious, sometimes more inspired but less controllable shadow remained. Plotinus, the third-century neo-Platonist, commented that 'feelings can be present without awareness of them'; and that 'the absence of a conscious perception is no proof of the absence of mental activity'. A century later, St Augustine famously wrote: 'I cannot totally grasp all that I am. The mind is not large enough to contain itself: but where can that part of it be which it does not contain?' Aquinas, in the thirteenth century, noted that 'there are processes in the soul of which we are not immediately aware'.[3]

Shakespeare clearly recognised the whole variety of unconscious influences on conscious life. He touches on the inability to see the source of one's own experience, or to comprehend its true meaning, in many of the plays, most famously, perhaps, in Antonio's lament at the beginning of *The Merchant of Venice*:

> In sooth, I know not why I am so sad:
> It wearies me; you say it wearies you;
> But how I caught it, found it, or came by it,
> What stuff 'tis made of, whereof it is born,
> I am to learn;
> And such a want-wit sadness makes of me,
> That I have much ado to know myself.

In *The Comedy of Errors* he notes the power of subliminal influences on perception, when he speaks of:

> . . . jugglers that deceive the eye,
> Dark-working sorcerers that change the mind.

While in *A Midsummer Night's Dream* he sketches a theory of creativity that anticipates by 300 years the insights that we encountered in Chapter 4:

> . . . as imagination bodies forth
> The form of things unknown, the poet's pen
> Turns them to shapes, and gives to airy nothing
> A local habitation and a name.

Prior to the sixteenth century, people's sense of their own minds was both deeper and broader than it was later to become: deeper, in embracing with equanimity the existence of internal forces that were beyond their ken; and broader, in accepting, for the most part rather uncritically, external sources of knowledge and authority.

'Mind' was not such an individual possession; it was embedded in and distributed across society. Over the course of the next two hundred years, however, both of these facets of the mind were to change significantly. First, it became normal, rather than abnormal, for individuals to 'make up their own minds' about things. Even by the year 1600, according to Lancelot Law Whyte, 'the person thinking for himself ceased to be a social freak inhibited by his difference from others, and began to claim the opportunity to realise himself and to guide the community'. And in the seventeenth century 'we can recognise the germ of a new experience and a new way of living which in our own time has become a social commonplace: the existentialist complaint that there is no tradition which makes life bearable . . . From then onward every sensitive and vital young person had to make his own choice.'[4] By the eighteenth century, the inclination and the ability to think for oneself was becoming firmly accepted as the goal of development, and the essential characteristic of maturity.

This increasing tendency to look for authority not outwards, to the received traditions and mythologies of Church or state, but inwards, to one's own mental workings, was accompanied by the shrinkage of 'mind' to admit only conscious reason, and to deny the legitimacy or even the very existence, of aspects of the mind that were not open to introspection. One came to conclusions, one *knew*, by working things out for oneself, and the way one did that was by conscious, deliberate thinking. That was the only kind of 'thinking', the only cognitive activity, there was. This outrageous, but as it turned out compelling, claim is usually attributed to René Descartes who, as Jacques Maritain puts it, 'with his clear ideas divorced intelligence from mystery'; though in truth Descartes' contribution was to encapsulate and give powerful expression to an intellectual movement that already possessed considerable momentum. Descartes wrote to his friend Mersennes: 'nothing can be in me, that is, in my mind, of which I am not conscious; I have proved it in the *Meditations*', and his insistence that the intelligent unconscious simply did not exist carried such conviction that it seeped into the culture at large, and became 'common sense'. All intelligence is conscious, and conscious intelligence – reason – is who, essentially, I am.

By 1690, John Locke was simply stating the obvious when he said that 'It is altogether as intelligible to say that a body is extended without parts, as that anything thinks without being conscious of it.' And in his *Essay Concerning Human Understanding* he neatly encapsulates what was fast becoming psychological orthodoxy:

> [A person] is a thinking, intelligent being, that has reason and reflection, and can consider itself as itself, the same thinking thing, in different times and places, which it does only by that consciousness which is inseparable from thinking, and, as it seems to me, essential to it . . . Consciousness always accompanies thinking, and it is that which makes every one to be what he calls self.

The assumption that conscious reason was the core of human identity, and the highest achievement of mental evolution, fed the growth of empirical science and the plethora of technological miracles to which it gave rise. From technology, it was a short step to the effective cultural takeover which we see today, and which Neil Postman, as we have seen, refers to as 'technopoly': a world in which every adverse or inconvenient circumstance is construed as a technological problem to be fixed through purposeful, rational analysis and invention.[5] In such a world, cognition becomes synonymous with this kind of busy, intentional mental activity – d-mode – and the very idea of 'letting things come', of waiting, becomes paradoxical or ridiculous. Thinking *is* the conscious manipulation of information and ideas preferably (now) with the aid of spreadsheets and graphics; and if solutions do not come, that simply means you are not thinking hard or clearly enough, or you need better data. Just as the invention of the printing press created 'prose', but by the same token made poetry appear exotic and élitist, and the balladeers redundant, so the hegemony of modern information technology has vastly increased the speed and complexity of data-handling, while making rumination and contemplation look hopelessly inefficient and old-fashioned.

Tools are not ideologically or psychologically neutral. Their very existence channels the development of intelligence (as well as a host of other facets of culture, such as vocational prestige), opening up and encouraging certain cognitive avenues, and simultaneously closing down and devaluing others. We are fashioned by our tools, and none more so than the computer. For 'the computer redefines people as "information processors" and nature itself as information to be processed. The fundamental metaphorical message of the computer, in short, is that we are machines – thinking machines, to be sure, but machines nonetheless.'[6] Computers epitomise the definition of 'intelligence' as fast, explicit and clear-cut, based on objective data and under tight control.

> The computer makes possible the fulfilment of Descartes' dream of the mathematization of the world. Computers make it easy to convert facts into statistics and problems into equations. And whereas this can be useful . . . it is diversionary and dangerous when applied indiscriminately to human affairs. So is the computer's emphasis on speed and especially its capacity to generate and store unprecedented quantities of information. In specialised contexts, the value of calculation, speed and voluminous information may go uncontested. But the 'message' of computer technology is comprehensive and domineering. The computer argues, to put it baldly, that the most serious problems confronting us at both personal and public levels require technical solutions through fast access to information otherwise unavailable.[7]

This implication is both false and pernicious. Our most serious problems are not technical, nor do they arise from inadequate information. If a war breaks out, or a family falls apart, it is not (usually) because of inadequate information. Yet to speak up for the virtues of mental sloth, for thinking that is elliptical and allusive, where even the goals may not be clearly known, is to be an information technology heretic. As they say in the trade, 'it does not compute'. Computers know nothing of the value of confusion, or the virtues of torpor. Their quality is assessed in terms of the size of their memory and the speed of their processor.

It is not the computer that is at fault so much as the computational frame of mind. Martin Heidegger, in his famous speech commemorating the birth of the German composer Conradin Kreutzer in 1955, drew a distinction between 'calculative' and 'meditative' thinking, and expounded powerfully the risks of leaning exclusively on the former.[8]

> Man today will say – and quite rightly – that there were at no time such far-reaching plans, so many inquiries in so many areas, [so much] research carried on as passionately as today. Of course. And this display of ingenuity and deliberation has its own great usefulness. Such thought remains indispensable. But – it also remains true that it is thinking of a special kind. Its peculiarity consists in the fact that whenever we plan, research and organize, we always . . . take [conditions] into account with the calculated intention of their serving specific purposes . . . Such thinking remains calculation even if it

neither works with numbers nor uses an adding machine or computer.

> Man finds himself in a perilous position . . . A far greater danger threatens [than the outbreak of a third world war]: the approaching tide of technological revolution in the atomic age could so captivate, bewitch, dazzle and beguile man that *calculative thinking may someday come to be accepted and practised as the only way of thinking*. What great danger then might move upon us? Then there might go hand in hand with the greatest ingenuity in calculative planning and inventing, indifference toward 'meditative' thinking, total thoughtlessness. And then? Then man would have denied and thrown away his own special nature – that he is a meditative being. Therefore the issue is the saving of man's essential nature. Therefore the issue is keeping meditative thinking alive. (Emphasis added)

The mind itself comes to be evaluated in calculative terms. Despite the tentative introduction of project and course work, students continue to be largely assessed, at school, college and university, on their ability to manipulate data under pressure of time. The so-called GMAT, the Graduate Management Admission Test, which is almost universally used in the States, and increasingly elsewhere, to select graduates for admission to management and business schools, consists of nine sections, seven of which contain a sequence of multiple-choice questions that is too long to be completed in the carefully stipulated time. The questions are designed to test 'basic mathematical skills and understanding of elementary concepts, the ability to reason quantitatively, solve quantitative problems, and interpret graphic data; the ability to understand and evaluate what is read; and the ability to think critically and communicate complex ideas through writing'.[9]

There is no reason to believe that these skills are anything other than useful or even vital for the manager of the future; but the implicit assumption that they cover *all* the important abilities seems, in the light of the research which I have been describing, staggeringly myopic. On the contrary, the ability to be innovative, or to detect the meaning in a snippet of information (the beginning of a consumer trend, for example) – abilities which companies frequently claim are in desperately short supply – requires expertise in slow and hazy, rather than fast and clear, ways of knowing. The GMAT seems designed to discard right from the start people with contemplative or aesthetic dispositions. Those with the potential and the inclination to become virtuoso ruminators need not apply. In general, all such tests of 'general ability', or 'intelligence' in the narrow sense, favour those who are able to think (a) fast, (b) under pressure, (c) on their own, about (d) abstract, impersonal problems, which are (e) clearly defined, have (f) a single 'right' answer, and have (g) been formulated by unknown other people.[10] Sometimes this ability is just what a practical situation demands. Many predicaments need something quite different. It is not surprising that tests of IQ correlate so poorly with measures of real-life, on-the-job performance. (As we saw in Chapter 2 the reasoning of professional horse-racing handicappers is based on a highly sophisticated (but largely unarticulated) mental model that includes up to seven independent variables – yet their individual levels of performance are completely unrelated to their IQs.[11])

In the business world particularly, the idea that the quality of thought depends on the amount and up-to-dateness of such information has completely taken over. It is all but impossible to resist the prevailing idea that all thinking worth its salt involves the deliberate manipulation of data – which boils down to 'facts and figures'. Everything is purposeful, explicit, calculated and well informed. Tom Peters, one of the most revered of management gurus, quotes, with apparent approval, the following.

> *Gentry* magazine, June–July 1994. A multipage advertisement for Silicon Valley realtor Alain Pinel includes, yes, each agent's Internet address. For example, you can reach Mary S. Gullixson at mgullixs@apr.com. One friend, who works for the firm, tells me she arrives at work each morning to about 100 e-mail messages![12]

> When Harriet Donnelly is on a business trip, she religiously carts along her NCR Safari 3170 notebook computer, her SkyWord alphanumeric pager and her AT&T cellular phone. She told *Fortune* magazine . . . that after a day of meetings: 'The first thing I do when I get back to my hotel is . . . return any messages I can, using voice mail. Next I plug my computer into the telephone and download [my] e-mail . . . I also get messages on my pager.'[13]

> One morning I stood on a wobbling dock [in Bangkok] waiting for a sooty old commuter 'express' boat to arrive. Next to me was a crisply starched Thai businessman . . . As he skilfully shifted his weight to keep his balance, he placed a string of

> calls on his portable cellular phone, talking excitedly. Wherever I went, I saw that kind of hustle, that kind of entrepreneurship, that kind of fervour . . . Frankly, it felt as if America had opted out of any effort to compete.[14]

If even Mary Gullixson and Harriet Donnelly – whose habits have rapidly become the business norm – aren't 'competing', it is clearly because American business is not quite 'up with the play' in terms of speed and information. The possibility that work depends on ideas, that ideas differ in their *quality*, as well as their up-to-dateness, and that quality takes time to mature, seems to be almost universally dismissed.

There are grumblings in the business world about the perils of rampant d-mode, though no one yet seems to see how deeply the problem runs. Roy Rowan, in his book *The Intuitive Manager*, talks about a business type which he calls 'the articulate incompetent': full of good ideas, immaculately presented, which lack substance and don't work. The verbal fluency of the articulate incompetents makes them persuasive; they are clever enough to be able to make an impressive-sounding case for whatever they have come up with. They tend to build an imposing superstructure of justification on a minimal foundation of observation. (And they may be so committed to being 'right' that they refuse to give their position up, even when it becomes apparent that there are considerations which they have not taken into account.) Rowan quotes interviews with some CEOs who tend to blame the phenomenon of articulate incompetence on schools. The inadequacy of the GMAT, and the 'education' that builds upon it, are noted by Robert Bernstein, chairman of giant publishers Random House, for example. Bernstein says: 'That's what frightens me about business schools. They train their students to sound wonderful. But it's necessary to find out if there's judgement behind their language.'[15]

Alternatively, there is a negative type, the 'articulate sceptic', whose cleverness manifests itself as a reflex need to show how bright he is by criticising whatever anyone else has proposed. As Edward de Bono has pointed out: 'The critical use of intelligence is always more immediately satisfying than the constructive use. To prove someone else wrong gives you instant achievement and superiority. To agree makes you seem superfluous and a sycophant. To put forward an idea puts you at the mercy of those on whom you depend for evaluation of the idea.'[16] It may be safer, for the essentially evaluative d-moder, to be seen to be reactive rather than proactive

– to respond to a presented problem, rather than to take a fresh look at a situation and reconceptualise what the problems are. Being generative, which is creative and intuitive, is bound to be riskier than being evaluative.

The need for creative responsiveness to changing conditions is now widely recognised in the pressurised cabins of business boardrooms. Grand strategic plans are fine in a stable world but, as Rowan says, talking of the senior manager: 'the farther into an unpredictable future his decisions reach . . . the more he must rely on intuition.'[17] As we have seen, it is not that in uncertain conditions we have to 'make do' with intuition, as if we were clutching at straws. It is that well-developed, tentatively used intuition is actually the best tool for the job; while the apparent solidarity of a rational, strategic plan offers nothing more than a comforting illusion. Henry Mintzberg, professor of management at McGill University in Canada, in his classic *The Rise and Fall of Strategic Planning*, demonstrates once and for all the insufficiency of d-mode as a way of knowing for the business world.[18] 'A good deal of corporate planning . . . is like a ritual rain-dance. It has no effect on the weather that follows, but those who engage in it think it does . . . Moreover, much of the advice related to corporate planning is directed at improving the dancing, not the weather.'[19]

Not only does inflexible attachment to a plan (which it has taken a lot of time, effort and money to create) make a company unresponsive; such plans, Mintzberg shows, tend to be based on only those considerations that can be clearly articulated and – preferably – quantified: 'hard data'. They therefore fail to take into account precisely that 'marginal' information – impressions, details, hunches, 'telling incidents' and so on – which provide the vital 'straws in the wind' on which prescient decisions can be based – and on which intuition thrives. Because consciousness demands information that is tidy and unequivocal, it can never be as *richly* informed as intuition. If you wait until a market trend is clear, you will have lost the edge. In business, as elsewhere, when decisions depend on the use of faint clues in intricate situations, the tortoise outstrips the hare.

But the calibre of intuition varies enormously. If intuition is merely a panicky, impulsive reaction to the failures of d-mode, it will be unreliable. In a world where TQM stands, more often than not, for 'Terrible Quality of Mind', intuition needs cultivating and nurturing. To be positive, we need to give some attention to specifying those conditions which facilitate the production of top-quality intuitions. The first requirement is a climate in which the value of intuitions, and the nature of the mental modes that produce them, are clearly understood by all. The second is leadership which models and acknowledges the value of intuition: managers who 'walk their talk' as far as slow knowing is concerned, encouraging the contribution of ideas that are judged on their merits, and not on how slickly or persuasively they are initially put across. Business leaders need to be open to 'the germ of an idea' that may seem unconventional at first sight, or which may be expressed in terms of analogies or images.

De Bono, in his many books on 'lateral thinking', has provided a wealth of telling illustrations of ideas that look 'silly' or 'childish' to start with, but which turn out on closer inspection to contain the seeds of highly creative and appropriate solutions.[20] A child drew a wheelbarrow that had the wheel at the back, near the handles, and the legs at the front. This 'silly mistake', which could easily have been 'corrected' by a conventional teacher, in fact produces a barrow that is ideal for manoeuvring round tight corners (on the same principle as a dumper truck that has the steering wheels at the back).

An apocryphal firm that sold mail-order glassware was suffering an unacceptably high level of breakages in transit, and they could not find a way to construct or label their parcels which made any difference. However much padding they put in, the glass was still getting broken. Then someone suggested that they simply stick the address label on the glass, and send it completely unprotected. Though this was rejected as being too risky, the idea behind it – that if the postal handlers could clearly *see* that the goods were fragile, rather than being told so by stickers which they tended to ignore, they might naturally take more care – was a good one, and it led, via a second creative leap, to the design of packaging that had a glassy finish, which had the desired effect.

A creative workplace needs to encourage people to engage with their work mindfully and to think about what they are doing. The development of such a working environment is stimulated by giving individuals, and especially teams, genuine responsibility for planning and carrying out meaningful pieces of work, and for deciding how their goals are to be best achieved. A recent official report on *Fostering Innovation* by the British Psychological Society, having surveyed all the relevant research, concludes that: 'Individuals are more likely to innovate where they have sufficient autonomy and control over their work to be able to try out new and improved ways of doing things' and where 'team members participate in the setting of objectives'.[21] The more people feel that they have some stake in their work, the more likely they are to be interested in spontaneously looking for improvements, and in keeping their thoughts, impressions and ideas simmering quietly on the back burners of their minds, both when they are 'on the job' and off it.

We saw in Chapter 5 that some daily routines and physical surroundings are more conducive to the germination of ideas than others. Giving workers some control over the work environment may be helpful – though individuals vary considerably in what works best for them, and many conducive conditions do not fit easily within the structure of conventional employment. I work best by the sea. Some people sink into the right frame of mind with the help of certain pieces of music. Often people are at their most receptive when they are able to spend a great deal of time in silence. Some people jog; some swim; some meditate. Descartes, it is said, did much of his best thinking lying in bed till late in the day. Encouraging people to note what works best for them is a practical first step. In a corporate setting it is not out of the question that – with a modicum of ingenuity – some of these conditions could be created, at least for some of the time.

People are obviously more inclined to be innovative and intuitive when they feel safe being so. *Fostering Innovation* suggests:

> One of the major threats to innovation is a sense of job insecurity and lack of safety at work . . . Where individuals are threatened they are likely to react defensively and unimaginatively . . . They will tend to stick to tried and tested routines rather than attempt new ways of dealing with their environments . . . People [are] generally more likely to take risks and try out new ways of doing things in circumstances where they feel relatively safe from threat as a consequence. It is the thrust of this document that the revolution in management practices in the 80s has now to be paralleled by a new revolution in the 90s and into the next century, which emphasises . . . psychological safety at work and practical support for the development and implementation of new and improved ways of doing things.

It is not just that people are bolder about trying things out when they feel relaxed and secure; threat creates a mindset of anxiety and entrenchment in which awareness is constricted and focused on the avoidance of the threat, rather than the spacious, open attitude that the slow ways of knowing require to work. People need to feel that they can say 'This may sound silly, but . . .'; or 'Can I just think aloud for a moment . . .' Where half-baked ideas are immediately torn to shreds, people rapidly learn to wait until d-mode has delivered a position that is polished and watertight (but quite possibly over-cautious and already out of date) – or not to contribute at all.

Finally there is time itself. The slow ways of knowing will not deliver their delicate produce when the mind is in a hurry. In a state of continual urgency and harassment, the brain-mind's activity is condemned to follow its familiar channels. Only when it is meandering can it spread and puddle, gently finding out such uncharted fissures and runnels as may exist. Yet thinking slowly, paradoxically, does not have to take a long time. It is a knack that can be acquired and practised. The mind needs to be *given* time; but its ingenuity also depends on the cultivation of a disposition to *take* one's time, as much as there is. One can learn to access and use these other ways of knowing more fluently. One might even suggest that managers – and their workforces – might try meditation; though, as a preliminary, they would need to understand what that means, and how it helps.

However, those who try to manage nations and corporations – ministers and executives of all persuasions – may be panicked by the escalating complexity of the situations they are attempting to control into assuming that time is the one thing they have not got. Their fallacy is to suppose that the faster things are changing, the faster and more earnestly one has to think. Under this kind of pressure, d-mode may be driven to adopt one shallow nostrum, one fashionable idea after another, each turning out to have promised more than it was capable of delivering. Businesses are re-engineered, hierarchies are flattened, organisations try to turn themselves into *learning* organisations, companies become 'virtual'. Meetings proliferate; the working day expands; time gets shorter. So much time is spent processing information, solving problems and meeting deadlines that there is none left in which to think. Even 'intuitive thinking' itself can easily become yet another fad that fails – because the underlying mindset hasn't changed.

Although it is important to think about how to encourage the *appearance* of slow knowing, it is even more important to think about the conditions that equip people with the longer-term *dispositions*, the *personal qualities* and the *capabilities* to make full use of their varied ways of knowing, regardless of the messages of the particular setting they happen to find themselves in. Even when the culture is implicitly directing them to use d-mode, people need to know how to make use of slow knowing, and when it is appropriate. This, at the current point in history, must surely be the true function of education.

In any school or college, there is not one curriculum but two. The first we might call the *content curriculum*: it is the body of knowledge and know-how that people are there to learn – sums, French, philosophy, dentistry, whatever. Both students and teachers are, all being well, clear about what the subject is and how progress is to be gauged. If this were the only curriculum, teachers would be free to use whatever means they could to make learning easier, quicker, more pleasant and more successful. But it isn't. Underneath every concern with content lies another curriculum, less visible but just as vital – the *learning curriculum* – which is teaching students about learning itself: what it is; how to do it; what counts as effective or appropriate ways to learn; what they, the students, are like as learners; what they are good at and what they are not. And if US Labor Secretary Doug Ross and 'Campaign for Learning' chairman Sir Christopher Ball are right, and the future of both social and personal wellbeing depends on people's confidence and competence *as learners*, then this second curriculum simply cannot be ignored. The learning society requires, above all, an educational system which equips all young people – not just the academically inclined – to deal well with uncertainty.

People come to any learning experience with a set of learning-related abilities and attitudes which will determine – in conjunction with the learning task and the learning culture which surrounds it – how learning goes. These learning skills and dispositions are themselves changed by experience: people's capabilities as learners, their fortitude in the face of difficulty, their implicit understanding of what learning entails, and their images of themselves as learners will all have been altered somewhat (even if only confirmed or consolidated) by any learning event. It is with the cumulative nature of such changes that the learning curriculum is concerned. It asks: how can this succession of learning challenges and encounters be designed and presented so that, over the long term, people's learning power changes in a positive direction?

The learning curriculum does not compete, or alternate, with the content curriculum: it follows it like a shadow. To be concerned with the education of young people as learners, generically, does not mean that we give up teaching them specific subjects. Like all human qualities, those of the 'good learner' develop in the course of engaging in appropriate activities. There has to be some content, something to learn *about*. Questions about the content curriculum remain to be answered. But the criteria which are used to determine the selection of subject matter, the methods of teaching and learning that are encouraged, and the focus of assessment: these are altered by the acknowledgement of the second curriculum. Whatever the topic, part of educators' attention remains on the mastery of the specific skills and materials; but another part now has to rest on the long-term residue of learning dispositions and capabilities which is accruing from those encounters.

We cannot simply assume that what is good in terms of one curriculum is necessarily good in terms of the other. If we wanted a swimmer to improve on her personal best time, we could threaten her with dire consequences if she did not. We could even, if all else failed, tow her rapidly up and down the pool. Neither of these ploys, however, would do much for her long-term performance. The fear of failure would probably make her tense and hostile. The towing, if we persisted with it, would succeed only in weakening her muscles. There is direct evidence that the two curricula do sometimes shear apart in just this way.

Carol Dweck of the University of Illinois has explored in a detailed series of experiments the quality of 'good learning' that we might call *resilience*: the ability to tolerate the frustrations and difficulties that inevitably occur in the course of learning, without getting upset and withdrawing prematurely. Dweck found that people's resilience varies enormously, right across the whole educational age spectrum from pre-school to graduate-level study. Some people, when they hit difficulty, would quickly start to become distressed, and instead of persisting with the task would have to find ways of shoring up their self-esteem. For them, the experience of finding learning hard was aversive. Her results showed that, of all the students she tested, it was the 'bright girls', those who were doing relatively well on the content curriculum, who were most likely to show this lack of resilience: to be 'failing', in other words, on the learning curriculum. Dweck speculated that the bright girls are the group with the most fragile grasp on learning precisely because they are the students with the least experience of sticking with difficulty, and (sometimes) succeeding. Being 'bright', they find learning relatively easy, so they encounter difficulty more rarely than a less 'able' child. And being girls, they are likely, when they *do* get stuck – especially when they are young – to be comforted or offered an alternative activity by a teacher, whereas a boy is more likely, so the research shows, to be encouraged to 'stick with it'. Thus the bright girls are the students who have had the fewest opportunities to build up their 'learning muscles', and their stamina consequently remains weak.[22]

Any learners who lack resilience will be fine while learning is going smoothly, but will be prone to fall apart when it gets rocky, and this vulnerability leads them, Dweck discovered, to make conservative learning choices. Those lacking resilience will choose tasks that they know how to handle, and become anxious when a teacher changes the rules, or offers a different, less familiar kind of learning experience. They will, as a result, tend to develop only a narrow range of learning skills, those that offer the highest probability of success. If doing well in a school is defined in terms of the ability to articulate and explain, then only d-mode will be exercised and developed. The risk is that without resilience, some young people, the 'successes', will overdevelop one side of their learning potential and neglect the other; while a different set of young people, the relative 'failures', may hardly develop as learners at all, or even go backwards. At worst, there is an unenviable choice between stunting and lopsidedness.

The learning curriculum, therefore, must first and foremost be committed to the strengthening of resilience, and this requires conveying to young people an accurate view of the many faces of learning, of the mind, and of themselves. Learning sometimes involves confusion, for example, so it makes no sense to create a content curriculum which systematically deprives young people of the opportunity of getting used to being confused and of learning how to deal with it productively; nor to create, or allow to be created, a learning culture which implicitly construes confusion as aversive or presumes that learning can and/or should happen without confusion; nor one in which learners come to feel that their sense of self-worth is contingent on clarity, and subverted by ignorance. If children learn to feel threatened by ignorance, their resilience will be weakened; and likewise if they learn to feel threatened by failure or frustration.

Carol Dweck has shown that resilience is also undermined by a false view of 'intelligence'. She distinguished between two general views of intelligence or 'ability'. In one view – the more accurate one which I am using here – ability is seen as a kind of expandable toolkit of ways of learning and knowing. As one learns, so one can also be learning how to learn; becoming a better learner. The other, which unfortunately tends to permeate educational discourse, sees 'ability' as an innately determined endowment of general-purpose brainpower which places a ceiling on what you can expect, or be expected, to achieve. To say of a child 'Sally is tall, has brown eyes, and is very bright' is implicitly to be subscribing – and encouraging Sally to subscribe – to the latter view. In such a view, no amount of effort on Sally's part will change her height, her eye colour or her 'ability'.

Dweck's dramatic discovery is that a lack of learning resilience is frequently underpinned by this latter, deterministic view of the mind. A child who agrees with the idea that 'it is possible to get smarter' is likely to be persistent and adventurous in her learning. A child who disagrees, who thinks that ability is fixed, is more likely to get upset when she fails. Why? Because the fluid 'theory' of intelligence encourages a child to stretch herself: by doing so she might become cleverer. That is a possibility for her. For her less fortunate classmate, it is not. She feels herself to be possessed of a certain immutable amount of cleverness, so that failure or confusion, for her, can only be construed as evidence that her ability is inadequate. And the more discomfited she is by this, the more impelled she is to withdraw, hide, defend or attack. The practical lesson of this research for the learning curriculum is that teachers themselves must understand the fluid and variegated nature of learning ability, and use language that conveys this view to children.

Once this view of the mind as expandable is established in principle, the next demand of the learning curriculum is that it should offer the opportunity to practise the whole gamut of ways of knowing and learning. On the foundation of resilience can be built greater *resourcefulness*. D-mode must be developed and refined, but so must the powers of intuition and imagination, of careful, non-verbal observation, of listening to the body, of detecting (without harvesting them too quickly) small seeds of insight, of basking in the mythic world of dream and reverie, of being moved without knowing why. If they were more aware of both the possibility and the value of doing so, teachers would be able to find a host of opportunities to vary more widely the learning modes which their lessons encouraged or demanded, and this is especially true of teachers in secondary, further and higher education. Many of the ways of knowing which we have been exploring are familiar to the primary school child and her teacher. But conventional content curricula tend to make the mistake of seeing them as 'childish', to be supplanted, as quickly as possible, by more explicit, more articulate forms of cognition. This attitude is profoundly misguided. The slower ways of knowing do not need to be replaced. They need to be cultivated and nurtured, right on into adulthood; and they need to be supplemented, not overshadowed, by the more formal ways of knowing that start later.

Intuition, for example, can readily be honed by including it explicitly within the learning context. Recent research on the learning of science shows that children develop a much richer understanding of how to *do* science, a much firmer, more flexible grasp of scientific thinking, if they are encouraged to bring their intuitions about how the world works into the laboratory with them: to share them, explore them and test them out. As we saw in Chapter 4, intuition is a vital way of knowing in scientific research. By working with their intuitions rather than ignoring them, children are learning not just science as a body of knowledge, but to think like scientists.[23]

Cultivating a relaxed attitude of mind, in which one can 'let things come', is also something that education could address. Some young people have picked up the knack for themselves; others may need a little coaching. Archbishop William Temple was clearly one of the former.

> When I was a boy at school I used to be set the task of composing poetry in Latin, which was, as you know, rather difficult. However, I was working by candle-light, and whenever I got 'stuck' and couldn't find the right phrase, I would pull off a stick of wax from the side of the candle and push it back, gently, into the flame. And then the phrase would simply come to me.[24]

Likewise there is good evidence for the value of imagination as a learning tool throughout the lifespan. Whether it be in learning physical skills such as sports, in preparing oneself for difficult encounters, or in sorting out one's own values and beliefs, active imagination and visualisation can often prove much more effective than rational self-talk.[25] Imagination and fantasy are areas in which young children are naturally expert, and in which their learning power can easily be refined and developed as they grow up. Conversely, if their imaginative birthright is allowed to atrophy through progressive neglect, learning power will be narrowed and reduced.

What about mindfulness? If we want young people to become successful on the content curriculum, we can afford to teach them as if knowledge were certain. In this kind of teaching, it may well be more efficient to adopt the 'textbook' approach: to act as if knowledge, and the appropriate methods for dealing with it, were (for the most part) agreed and secure. 'Why risk unsettling children with fancy talk about the "social construction of knowledge"?' is a perfectly fair question. But on the learning curriculum students must be helped to develop a greater sense of ownership of knowledge and the knowledge-making process, and this means presenting knowledge as more equivocal; a human product that is always open to question and revision. If we want to start children out on a journey that heads, however remote the destination, in the direction of wisdom, then we may need to run the risk of creating some epistemological insecurity. In fact it turns out that the risk is not so great.

Ellen Langer at Harvard has conducted a series of studies with both high school and college students in which different groups were presented with the same information in different ways. For example, in one study undergraduates were given a paper to read which described a theory about the evolution of urban neighbourhoods. For one group of students the paper was written as if the theory were the simple truth. For another group, it was presented *as* a theory, using phrases such as 'You could look at the data this way', or 'It may be that . . .' When tested on their ability to use the knowledge they had gained, Langer found that, though retention was the same for the two groups, the 'could be' group was much better at using the information in flexible and creative ways. She concluded that the fear of making children insecure about knowledge was groundless, provided teachers present its provisional status as an intrinsic feature of knowledge, and not as personal indecisiveness. 'Children taught conditionally [in this way] are *more* secure, because they are better prepared for negative or unexpected outcomes.'[26]

How does the distinction between the content and the learning curricula bear on the old debate between traditional 'chalk-and-talk' teaching and 'discovery learning'? From the point of view of the content curriculum, it can seem terribly inefficient to allow children to flounder around 'reinventing the wheel', when there are so many different 'wheels' that have to be learnt. If the important thing is the wheels, this objection is entirely valid. But within the learning curriculum, what matters most is not the wheel but the inventing – and the strengthening of the powers of invention which occurs through being allowed and encouraged to invent. Time spent discovering things for yourself, even though someone could have simply told you the answer or given you the information, may be time well spent if the outcome is greater confidence and competence as an explorer. Discovery learning both draws upon and develops the power of 'learning by osmosis', and like intuition and imagination, this ability to extract patterns from experience, without necessarily being able to say what they are, continues to be of inestimable use throughout life. Both the acquisition of knowledge and skill *and* the development of learning power are important. Learners, whether children or adults, can flounder unproductively if they are given neither the tools nor the knowledge that they may need to get started on a piece of learning. The real enemy of the learning curriculum is dogmatism, whichever side it takes.

To be resourceful, in these terms, is to have at your disposal the full range of learning resources – different ways of knowing – and to have developed a good intuitive sense of the kinds of problem which each is good for, and the kinds of knowledge that each delivers. The resourceful learner is able to attend to puzzling situations with precision and concentration, and also with relaxed diffusion. She is able to 'let things speak', to see what is actually there, and not, as Hesse put it, to observe everything in 'a cloudy mirror of your own desire'. She is able to make good use of clues and hints. She is able to analyse and scrutinise, but also to daydream and ruminate. She is able to ask questions and collaborate, but is also able to keep silent and contemplate. She is able to be both literal and metaphorical, articulate and visionary, scientific and poetic: to know as Madame Curie, and to know as Emily Dickinson. To be a resourceful learner is to have had the opportunity to play, explore and experiment with each of these ways of knowing and learning, so that their power, their precision and their pertinence have all been uncovered.

The resourceful learner has also to develop the ability to be a good 'manager' of her own learning projects: to be able to judge when an approach to a problem appears not to be working, when to persist, when to change tack, and when to give up. Good learning requires the ability to be *reflective*; to take a strategic, as well as a tactical, perspective on one's learning and knowing; to be aware of 'how things are going', and of what alternative approaches there might be. Thus the learning curriculum demands that learners take on, at an appropriate rate, some genuine responsibility for deciding what, when and how they will learn, and for evaluating their own efforts. Knowing *how* to know develops through the discovery of the strengths, weaknesses and limitations of different learning styles and strategies, as they are applied across a range of real-life learning settings.

One cannot, as I have said, treat 'learning' as a new 'subject' to be added to the content curriculum. The evidence of 'study skills' programmes, for example, shows that learning strategies cannot successfully be taught directly, and that any benefits that do accrue tend not to transfer from explicit lessons into spontaneous use.[27] Learning power grows through experience; it cannot be reduced to formulae and transmitted into someone's head by instruction. Thus where the content curriculum might demand tight scheduling and supervision, the learning curriculum suggests that students be given some time, freedom and encouragement to explore. On the content curriculum, it is important that learners are told how well they are doing by being measured against 'objective' criteria: such feedback informs them of their progress, and may 'motivate' them if it doesn't demoralise them. On the learning curriculum, however, it is vital that learners are given some responsibility, encouragement and assistance to reflect upon the value of their own efforts, because only by doing so will they develop an intuitive 'nose' for quality; the ability to tell for themselves, in terms of their own (largely tacit, and certainly not quantifiable) values, what is 'good work'.

On the content curriculum, it may be seen as damaging if students are set problems that are too easy – they will be bored – or too hard – failure will dent their self-esteem. On the learning curriculum, there is less need to protect students from difficulty, or from 'biting off more than they can chew', for learning power is strengthened and broadened by the attempt to chew, and much of value may be learnt by pondering Eliot's 'Ash Wednesday' when you are ten (if you want to), just as it may from going fishing with an elder sister, even though you are too small to lift the rod, or from 'helping' your mother with the crossword, even though you solve no clues. If you are always fed a diet of problems that have been neatened up and graded, you are deprived of the opportunity to develop those slow, intuitive ways of knowing that are designed precisely to work best in situations that are untidy, foggy, ill conceived.

If the learning society is to evolve, practical changes to workplace ethos and educational methods, of the kind I have been sketching, need to be encouraged. But at a deeper level we are being asked nothing less than to conceive of the human mind in a new way. Descartes' legacy to the twentieth century is an image of the mind as 'the theatre of consciousness', a brightly illuminated stage on which the action of mental life takes place; or perhaps as a well-lit office in which sits an intelligent manager, coolly weighing evidence, making decisions, solving problems and issuing orders. In this executive den, human intelligence, consciousness and identity come together: they are, in effect, one and the same thing. 'I' am the manager. 'I' work in the light. I have access to all the files that comprise my 'intelligence'. What I cannot see, or see into, either does not exist, or it is mere 'matter', the dumb substance of the body that can do nothing of any interest on its own. It may manage certain menial operations like digestion, respiration and circulation without supervision; but to do anything clever it has to wait for instructions from head office.

This image continues to animate and channel our sense of our own psychology, of our potentialities and resources, and it is wrong in every regard. The naïve mind–body dualism on which it rests is philosophically bankrupt and scientifically discredited. *Unconscious* intelligence is a proven fact. The need to wait for inspiration rather than to manufacture it – to envisage the conscious self as the recipient of gifts from a workplace to which consciousness has no access – is likewise undeniable. We need now a new conception of the unconscious – one which gives it back its intelligence, and which reinstalls it within the sense of self – if we are to regain the ways of knowing with which it is associated. Highlighting the ways of knowing that are associated with consciousness, control and articulation enabled the extraordinary explosion of scientific thinking and technological achievement of the last two centuries; but the cost was a disabling of other faculties of mind that we cannot afford to be without. As Lancelot Law Whyte puts it:

> The European and Western ideal of the self-aware individual confronting destiny with his own indomitable will and skeptical reason as the only factors on which he can rely is perhaps the noblest aim which has yet been accepted by any community . . . But it has become evident that this ideal was a moral mistake and an intellectual error, for it has exaggerated the ethical, philosophical and scientific importance of the awareness of the individual. And one of the main factors exposing this inadequate ideal is the [re-]discovery of the unconscious mind. That is why *the idea of the unconscious is the supreme revolutionary conception of the modern age.*[28] (Emphasis added)

This conception of the unconscious – which I have been calling the undermind – is very different from the notions of the unconscious that twentieth-century European culture generally admits – such as the Freudian subconscious, the sump of the mind into which sink experiences, impulses and ideas too awful or dangerous to allow into consciousness. This representation of the unconscious is pathological and repressed. It accepts the basic Cartesian premise that consciousness is intelligent and controlled, and therefore the corollary that the unconscious must be other than, and opposed to, consciousness: emotional, irrational, wild and alien. The unconscious cannot be 'I'; it has to be 'it' – '*das Es*', as Freud originally called it, before it was gratuitously mystified by its translation into English as 'the Id'.

Clinical practice and the development of psychotherapy in the nineteenth and twentieth centuries have shown that we do indeed require this sense of the unconscious to explain aspects of human experience and behaviour – but the Cartesian image is left basically unchallenged if we make the mistake of assuming that this dark, subversive corner of the mind is the *only* part that lies outside conscious awareness. Even if we add on, as Arthur Koestler said of Jung, a kind of exotic 'mystical halo' to this fundamentally pathological picture, the core alliance of consciousness, intelligence and identity survives. The images of the unconscious that have resurfaced and survived in contemporary culture are merely elaborations of, or footnotes to, an image that continues to control the way we think about ourselves.

Yet throughout the last 350 years there *has* been a succession of other voices, demanding that the undermind be returned to its central place in our view of the mind. Less than twenty years after the publication of Descartes' *Meditation* we have Blaise Pascal reminding us that 'the heart has its reasons of which reason itself knows nothing'. Before the end of the seventeenth century Cambridge philosopher and scientist Ralph Cudworth was writing: 'It is certain that our human souls themselves are not always conscious of whatever they have in them; for even the sleeping geometrician hath, at that time, all his geometrical theorems some way in him. [And] we have all experience of our doing many . . . actions non-attendingly.' And Sir William Hamilton, one of the first philosophers writing in English to be influenced by the rise of the German Romantic movement, was lecturing on the proposition that 'The sphere of our conscious modifications is only a small circle in the centre of a far wider sphere of action and passion, of which we are conscious only through its effects.'

In 1870 the French historian and critic H. A. Taine wrote an essay, '*Sur l'Intelligence*', in which he deliberately elaborated the image of the 'Cartesian theatre' to include its unconscious background.

> One can therefore compare the mind of a man to a theatre of indefinite depth whose apron is very narrow but whose stage becomes larger away from the apron. On this lighted apron

> there is room for one actor only. He enters, gestures for a moment, and leaves; another arrives, then another, and so on . . . Among the scenery, and on the far-off backstage there are multitudes of obscure forms whom a summons can bring on to the stage or even before the lights of the apron, and unknown evolutions take place incessantly among this crowd of actors of every kind to furnish the stars who pass before our eyes one by one.

Here, finally, we may be able to turn the Cartesian image of the mind against itself, for the bare image of a 'thought in a spotlight' hardly does justice to our understanding of 'theatre'. It leaves out at least two things without which theatre simply isn't theatre: the wings, and the nature of drama itself. The action on the stage only makes sense in terms of entrances and exits. Actors are not borne on stage; they *arrive*, and then, after a while, they *leave*. And we know they arrive from somewhere, and go to somewhere. If we do not know – even when we are not consciously thinking about it – that there *is* a 'behind the scenes' of dressing-rooms, technicians, props and paraphernalia, and a hidden world of rehearsal and discussion in which interpretations and performances are much more fluid and tentative than those that finally appear, in costume, in front of the footlights, then we don't understand what 'theatre' is. The visible performance presupposes an enormous amount of invisible apparatus and activity.

Likewise we are liable to become very confused if we don't understand the distinction between the actor and the role; between drama and 'real life'. What we watch in the theatre is a simulation, a fiction, that is designed to resemble 'real life', but also to *dramatise* it; to distort, highlight, doctor and if necessary mislead, in order to make a point or create an effect. If you forget that what is happening on stage is not 'real', then you will find yourself cowering in your front-row seat when the villain pulls a gun, or scrambling up on to the stage in order to save the heroine from a fate worse than death. A good 'play' may engage your attention and your sympathies, you may 'lose yourself' in its world for a while, but if ultimately you cannot tell the difference between play and reality, you will be in trouble. To assume that consciousness is showing and telling us the complete and literal truth is to make precisely that mistake. Thus once we begin to analyse the metaphor of the theatre, it starts immediately to unravel and subvert itself. We find that we cannot do without the wings of the undermind; and we cannot take what is going on in consciousness at face value.

An image such as the expanded theatre can help to convey a feeling for the new relationship between conscious and unconscious that we are seeking, but, in a d-mode culture, such images do not carry much weight. The voices of philosophy, poetry and imagery are relatively weak in a world that largely assumes that only science and reason speak with true authority. Thus, paradoxically, it is only science itself that can bring credible tidings of unscientific ways of knowing. One must speak to d-mode in its own language if it is to entertain the idea that it may itself be limited. The empirical research on the slower ways of knowing, and on the cognitive capacity of the undermind, can contribute significantly to the creation of the much-needed shift in our understanding of the mind. As this research gathers further momentum, it will, it must be hoped, seep into the culture at large, and encourage educators, executives and politicians to use mental tools more suited to the intricate jobs that confront them. The hare brain has had a good run for its money. Now it is time to give the tortoise mind its due.

Notes

Chapter 1

1. See Fensham, P. J. and Marton, F., 'What has happened to intuition in science education?', *Research in Science Education*, Vol. 22 (1992), pp. 114–22. This paragraph comprises a collection of assertions which I shall unpack and justify as the argument unfolds.

2. Norberg-Hodge, Helena, *Ancient Futures: Learning from Ladakh* (Shaftesbury: Element, 1991).

3. Postman, Neil, *Technopoly* (New York: Knopf, 1992).

4. The demonstration that there is a source of 'unconscious intelligence' in the mind is crucial to the argument of the book, and so that we can talk about the processes and properties of this source, it needs a name – or names. Sometimes I shall just call it 'the unconscious', and distinguish it from the Freudian repository of repressed memories by referring to the latter as the 'subconscious'. Where the contrast needs making more strongly, I shall use the expressions 'the intelligent unconscious' or 'the cognitive unconscious'. For variety, and where I feel there may be a danger that using the term 'the unconscious' may import unhelpful connotations, I shall also use my own coinage, 'the undermind'. When we go on to discuss how unconscious intelligence is exemplified in and generated by real flesh-and-blood human beings, I shall refer to the source as 'the unconscious biocomputer', or sometimes just as the 'brain-mind'. My intention is that, by using a range of different terms, I shall be able to build up a composite picture of 'the intelligent unconscious' that does justice to its many faces and functions.

5. There is presently an explosion of interest, both popular and scholarly, in the subject of consciousness. Julian Jaynes' *The Origin of Consciousness in the Breakdown of the Bicameral Mind*, Daniel Dennett's *Consciousness Explained*, Roger Penrose's *Shadows of the Mind* and Robert Ornstein's *The Psychology of Consciousness* are just four of the dozens of books that have appeared in the last few years on the nature, the evolution and the function of the conscious mind. My book is partly a reflection of this wave of enthusiasm, and partly a reaction against it. I certainly have things to say about what consciousness is and what it is for, but I also argue that we cannot

understand the nature of the conscious mind without having a better image of the dark, inaccessible layers – the minds behind the mind – that underlie it, and from which it springs. Consciousness can only be understood in relation to the unconscious. If we persist in trying to make sense of consciousness in and on its own terms, we shall continue to see those modes of mind that are most associated with consciousness as pre-eminent; and continue to ignore or undervalue those that are less conscious, less deliberate, or which require a different image of mind if they are to become visible, and to make sense. Fascinating though it is, much of this wave of research and speculation on consciousness must be seen as symptomatic of our cultural obsession with the conscious intellect, and not a corrective to it. None of these books has anything to say about the practical effect on our psychology of this reconceptualisation of the relationship between conscious and unconscious.

6. It should be clear that what I am talking about here is not a thinly disguised version of the distinction between the 'right' and 'left' brain, which was popular a few years ago as a way of thinking about the brain's lost capacities. Though it persists in the literature of 'pop psychology', this distinction has run out of steam. First, in all but a few unfortunate people, the brain works as a whole. Functionally it does not have two separate halves. To ask people to switch from a 'left-brain' to a 'right-brain' way of thinking would be like insisting that people switched from driving an 'engine car' to a 'steering-wheel car'.

Secondly, people's aspirations for the 'right brain' ran wildly out of control: far ahead of what the scientific research could possibly justify. Certainly there is a greater linguistic ability, for most right-handed people, in the left cerebral hemisphere of the brain than in the right; but the research shows that there is language on the left side, just as there are 'holistic' properties on the right, as well. Michael Gazzaniga, one of the scientists, along with Nobel laureate Roger Sperry, whose work on so-called 'split-brain' patients was the inspiration of the 'left brain, right brain' distinction, wrote despairingly as long ago as 1985: 'How did some laboratory findings of limited generality get so outrageously misinterpreted? . . . The image of one part of the brain doing one thing and the other part something entirely different was there [in the popular literature], and [the fact] that it was a confused concept seemed to make no difference.' (Gazzaniga, M., *The Social Brain: Discovering the Networks of the Mind*, New York: Basic Books, 1985).

When I talk here about 'tortoise mind', the 'undermind' or the 'intelligent unconscious' I am not talking about a new *bit* of the brain. I am talking about a set of different modes of mind that, above all, require a less busy, less purposeful, less problem-orientated mental ambience.

Chapter 2

1. Gardner, Howard, 'The theory of multiple intelligences'. Presentation to the Annual Conference of the British Psychological Society Division of Educational and Child Psychologists, York, (January 1996).

2. Goleman, Daniel, *Emotional Intelligence* (New York: Bantam, 1995).

3. Rozin, Paul, 'The evolution of intelligence and access to the cognitive unconscious', in Sprague, J. M., and Epstein, A. N. (eds), *Progress in Psychobiology and Physiological Psychology*, Vol. 6 (New York: Academic Press, 1976). Rozin was one of the first to use the term 'cognitive unconscious', and I have borrowed other arguments and examples in this section from his seminal paper.

4. See Wooldridge, D., *The Machinery of the Brain* (New York: McGraw Hill, 1963).

5. Studies by Aronson (1951), reported by Rozin, op cit, p. 252.

6. See Smith, Ronald, Sarason, Irwin and Sarason, Barbara, *Psychology: the Frontiers of Behavior*, 2nd edition (San Francisco: Harper and Row, 1982), p. 273.

7. The research that confirms these differences has been recently summarised by Reber, Arthur, *Implicit Learning and Tacit Knowledge: an Essay on the Cognitive Unconscious* (Oxford: OUP, 1993).

8. Some of the studies that lead to this conclusion will be discussed in Chapter 8.

9. Carraher, T. N., Carraher, D. and Schliemann, A. D., 'Mathematics in the street and in schools', *British Journal of Developmental Psychology*, Vol. 3 (1985), pp. 21–9. Ceci, S. J. and Liker, J., 'A day at the races: a study of IQ, expertise and cognitive complexity', *Journal of Experimental Psychology: General*, Vol. 115 (1986), pp. 255–66.

10. Berry, Dianne C. and Broadbent, Donald E., 'On the relationship between task performance and associated verbalizable knowledge', *Quarterly Journal of Experimental Psychology*, Vol. 36A (1984), pp. 209–31. See also the recent overview of these results in Berry, Dianne and Dienes, Zoltan, *Implicit Learning* (London: Lawrence Erlbaum Associates, 1992).

11. Watzlawick, Paul, Weakland, John and Fisch, Richard, *Change: Principles of Problem Formation and Problem Resolution* (New York: Norton, 1974).

12. A great body of this work is summarised in Lewicki P, Hill, T. and Czyzewska, M. 'Nonconscious acquisition of information', *American Psychologist*, Vol. 47 (1992), pp. 796–801.

13. Reber, op cit.

Chapter 3

1. Developmental psychologist Annette Karmiloff-Smith makes the same observation in her book *Beyond Modularity: a Developmental Perspective on Cognitive Science* (Cambridge, MA: MIT, 1992).

2. These studies are comprehensively reviewed in Berry and Dienes, op cit.

3. In fact there is research to show that our faith in articulation, as a measure of competence, is misguided in the 'real world'. Medical students' performance in written examinations, for example, is quite unrelated to their clinical skill and judgement. Yet our implicit (in the sense, this time, of 'unquestioned') faith in good old exams is such that we continue to put generations of students through them. See Skernberg, R. T., and Wagner, R. K. (eds), *Mind in Context* (Cambridge: CUP, 1994).

4. See Wason, Peter, and Johnson-Laird, Philip, *The Psychology of Reasoning: Structure and Content* (London: Batsford, 1972).

5. Coulson, Mark, 'The cognitive function of confusion', paper presented to the British Psychological Conference, London (December 1995).

6. Lewicki *et al*, op cit.

7. Master, R. S. W., 'Knowledge, knerves and know-how: the role of explicit versus implicit knowledge in the breakdown of a complex skill under pressure', *British Journal of Psychology*, Vol. 83 (1992), pp. 343–58.

8. The polar planimeter, as a metaphor for know-how, is described by Runeson, Sverker, 'On the possibility of "smart" perceptual mechanisms', *Scandinavian Journal of Psychology*, Vol. 18 (1977), pp. 172–9.

9. Bourdieu, Pierre, *In Other Words: Essays towards a Reflexive Sociology* (Stanford, CA: Stanford University Press, 1990).

10. Huxley, Aldous, *Island* (London: Chatto, 1962).

11. Korzybski was the founder of the intellectual movement known as 'general semantics', influential in the 1940s and 1950s, that explored the relationship between language and human experience: see, for instance, his book *Science and Sanity* (New York: W.W. Norton, 1949).

Chapter 4

1. From Spencer, Herbert, *An Autobiography*, quoted in Ghiselin, Brewster (ed), *The Creative Process* (Berkeley, CA: University of California Press, 1952).

2. This sense of 'intuition' is associated with a long tradition of religious

and philosophical writers, which includes pre-eminently Spinoza, and later Henri Bergson, who see intuition as being the royal road to some higher or 'spiritual' truth. In Spinoza's use of the term, intuition refers to a kind of profound, unmediated understanding of the 'nature of things' which arises through a deep contemplative communion with people and objects. This 'intuition', according to Spinoza, is inevitably accurate; it carries with it an unquestionable certainty and authority, and only makes its appearance after the purposeful questings of reason have exhausted themselves.

3. All these first three problems can be solved by filling the largest jar, and from it filling the middle-sized jar once and the smallest jar twice. You are then left with the desired volume in the big one. For an account of these studies, see Rokeach, Milton, 'The effect of perception time upon the rigidity and concreteness of thinking', *Journal of Experimental Psychology*, Vol. 40 (1950), pp. 206–16.

4. All you have to remember from your school maths is that the circumference of a circle is 6.28 times as big as the radius (i.e. twice the product of the radius and the constant 'pi', which is 3.14).
Suppose the radius of the smoothed-out Earth is R.
Then the original length of string is 6.28 times R.
If the size of the gap we are interested in is called 'r', then the new total radius is R+r, and the new total length of string is therefore 6.28 times (R+r).
But this is the original length, 6.28R, plus 2m, or 200cm. So

$$6.28(R+r) = 6.28R + 200.$$

Take away the = 6.28R from both sides of the equation, and divide both sides by 6.28. This leaves

$$r = 200/6.28, \text{ or about } 32\text{cm}.$$

5. This tension between reason and intuition has been known since antiquity. It is recorded, for example, that the Athenian general Nicias, at the siege of Syracuse, decided to follow an intuitive interpretation of a lunar eclipse and, against his 'better' – i.e. rational – judgement, to postpone a tactical retreat, his faith in intuition leading to a decisive defeat.

6. See, for example, McCloskey, M., 'Intuitive physics', *Scientific American*, Vol. 248 (1983), pp. 114–22.

7. Ceci, S. J. and Bronfenbrenner, U., 'Don't forget to take the cupcakes out of the oven: strategic time-monitoring, prospective memory and context', *Child Development*, Vol. 56 (1985), pp. 175–90.

8. This example actually relies on 'learning by osmosis' rather than the kind of intuition that is the main focus of this chapter. But it makes graphically a point which applies to both sorts of slow knowing.

9. Kahneman, Daniel and Tversky, Amos, 'Intuitive prediction: biases and

corrective procedures', in Kahneman, D., Slovic, P. and Tversky, A. (eds), *Judgement under Uncertainty: Heuristics and Biases* (Cambridge: CUP, 1982).

10. These quotations come from a survey carried out in 1992, in which eighty-three Nobel laureates in science – physics, chemistry and medicine – answered the question: 'Do you believe in scientific intuition?' The vast majority were in no doubt that a vital stage on the road to their discoveries was listening to hunches about their results, or promptings about the direction to take, which were quite incapable of rational defence or explanation. See Fensham, Peter, and Marton, Ference, 'What has happened to intuition in science education?', *Research in Science Education*, Vol. 22 (1992), pp. 114–22.

11. Spencer Brown, George, *Laws of Form* (London: Allen and Unwin, 1969).

12. Noddings, Nel and Shore, Paul, *Awakening the Inner Eye: Intuition in Education* (New York: Teachers' College Press, 1984).

13. Lowe, John Livingston, *The Road to Xanadu* (Boston: Houghton Mifflin, 1927).

14. Quoted by Gerard in Ghiselin, op cit.

15. Coleridge, Samuel Taylor, 'Prefatory note to Kubla Kahn', quoted in Ghiselin, op cit.

16. Quotation from Woodworth, R. S. and Schlosberg, H., *Experimental Psychology* (1954), quoted in Smith, S. M. and Blankenship, S. E., 'Incubation and the persistence of fixation in problem-solving', *American Journal of Psychology*, Vol. 104 (1991), pp. 61–87. See also Smith, S. M., Brown, J. M. and Balfour, S. P., 'TOTimals: a controlled experimental method for studying tip-of-tongue states', *Bulletin of the Psychonomic Society*, Vol. 29 (1991), pp. 445–7; and Smith, S. M., 'Fixation, incubation and insight in memory and creative thinking', in Smith, S. M., Ward, T. B. and Finke, R. A. (eds), *The Creative Cognition Approach* (Cambridge, MA: Bradford/MIT Press, 1995).

17. Yaniv, I. and Meyer, D. E., 'Activation and metacognition of inaccessible stored information: potential bases for incubation effects in problem-solving', *Journal of Experimental Psychology: Learning, Memory and Cognition*, Vol. 13 (1987), pp. 187–205.

18. These studies are reported in Bowers, K. S., Regehr, G., Balthazard, C. and Parker, K., 'Intuition in the context of discovery', *Cognitive Psychology*, Vol. 22 (1990), pp. 72–110; and Bowers, K. S., Farvolden P. and Mermigis, L., 'Intuitive antecedents of insight', in Smith, S. M. *et al* (eds), *The Creative Cognition Approach*, op cit. The solutions to the picture puzzles are: 1A shows a camera; 2A shows a camel.

19. The words in 1A are all associates of 'candle'; the words in 2B are all associates of 'carpet'; the words in 3A are all associates of 'pipe'.

20. The common associate to all fifteen words is 'fruit'. If you feel inclined to argue with some of these associations, bear in mind that they were all statistically defined as low-frequency associative responses to the cue word from a large-scale survey of American students.

Chapter 5

1. Skinner, B. F., 'On "Having" a Poem', reprinted in *Cumulative Record* (New York: Appleton-Century-Croft, 1972).

2. Quoted in Ghiselin, op cit.

3. These last two examples were used to illustrate a television programme, broadcast on 1 September 1996 on Channel 4, called 'Break the Science Barrier with Richard Dawkins'.

4. James, Henry, 'Preface to *The Spoils of Poynton*', in Ghiselin, op cit.

5. Canfield, Dorothy, 'How Flint and Fire Started and Grew', in Ghiselin, op cit.

6. Simonton, D. K., *Genius, Creativity and Leadership: Historiometric Inquiries* (Cambridge, MA: Harvard University Press, 1984).

7. The example is discussed by Schooler, Jonathan and Melcher, Joseph, 'The ineffability of insight', in Smith, Stephen *et al* (eds), *The Creative Cognition Approach* op cit. A general survey of the evidence of an 'inverted U'-shaped relationship between knowledgeability and creativity is provided by Simonton, op cit.

8. Westcott's 'successful intuitives' are similar to the character type identified by C. G. Jung as 'introverted intuitive'. These people are not only socially somewhat introverted, as Westcott's intuitives were; they are also those who, according to Jung, have the closest relationship with their own unconscious. See, for example, Jung's classic *Psychological Types*, translated by H. G. Baynes (London: Routledge, 1926). For Jung, intuition is one of four basic mental functions, the other three being 'thinking', 'feeling' and 'sensing'. It is a rather nebulous faculty which detects possibilities and implications in a holistic fashion, at the expense of details. Jung's view is that every person has one of these four modes which is more developed than, and used in preference to, the others. One has a disposition to be a thinking or an intuitive 'type', for example. In addition to these four functions, Jung proposed that we also differ in the extent to which our basic orientation is towards the external or the internal world – whether one is an 'extravert' or an 'introvert'.

Jung, as is well known, also distinguishes between two layers of the unconscious mind, the personal and the collective. In the personal unconscious lie those memories and perceptions which are too weak ever to make it over the borderline into consciousness, or which have been repressed. The collective unconscious contains the *archetypes*, forms of universal human-species knowledge derived from one's whole ancestral lineage, and reborn in each individual's brain structure. The collective unconscious makes itself known through our innate understanding of ubiquitous human situations and relationships, and through universal systems of symbols. 'Introverted intuitives', in Jung's scheme, do, in a sense, have access to a 'higher' knowledge than the other types by virtue of their more intimate relationship with the symbolic life and fundamental knowings of the collective unconscious.

Jung's typology (and the variety of personality tests, such as the Myers-Briggs Type Inventory, based upon it) now looks rather coarse in the light of our better, more empirically based, understanding of the way in which the unconscious itself generates intuitions. (Jung held to a view of intuition as a way of seeing *into* the unconscious, rather than as a product *of* it.) And there is a second important way in which Jung's pioneering work has been superseded. He, rather fatalistically, tended to talk of the four basic personality types as if they were constitutional, and therefore largely unalterable. We now know that the intuitive way of knowing is educable, capable of being enhanced and sharpened. It is for these reasons that Jung's ideas receive less attention in this book than some readers might have expected.

9. Westcott, Malcolm, *Toward a Contemporary Psychology of Intuition* (New York: Holt, Rinehart & Winston, 1968).

10. Schon, Donald, *The Reflective Practitioner: How Professionals Think in Action* (New York: Basic Books, 1983).

11. Rokeach, op cit.

12. Cowen, Emory L., 'The influence of varying degrees of psychological stress on problem-solving rigidity', *Journal of Abnormal and Social Psychology*, Vol. 47 (1952), pp. 512–19.

13. Combs, Arthur and Taylor, Charles, 'The effect of perception of mild degrees of threat on performance', *Journal of Abnormal and Social Psychology*, Vol. 47 (1952), pp. 420–4.

14. Kruglansky, A. W. and Freund, T., 'The freezing and unfreezing of lay inferences: effects on impressional primacy, ethnic stereotyping and numerical anchoring', *Journal of Experimental Social Psychology*, Vol. 19 (1983), pp. 448–68.

15. Wright, Morgan, 'A study of anxiety in a general hospital setting', *Canadian Journal of Psychology*, Vol. 8 (1954), pp. 195–203.

16. Prince, George, 'Creativity, self and power', in Taylor, I. A. and Getzels, J. W. (eds), *Perspectives in Creativity* (Chicago: Aldine, 1975).

17. Fischbein, Efraim, *Intuition in Science and Mathematics* (Dordrecht, Holland: Kluwer, 1987), p. 198.

18. Viesti, Carl, 'Effect of monetary rewards on an insight learning task', *Psychonomic Science*, Vol. 23 (1971), pp. 181–3.

19. Bruner, Jerome, Matter, Jean and Papanek, Miriam, 'Breadth of learning as a function of drive level and mechanisation', *Psychological Review*, Vol. 62 (1955), pp. 1–10.

20. Hughes, Ted, *Poetry in the Making* (London: Faber, 1967).

21. Emerson, R. W., 'Self-reliance', in *The Collected Works of Ralph Waldo Emerson*, Vol. II (Cambridge, MA: Belknap Press, 1979).

22. Lynn, Steven and Rhue, Judith, 'The fantasy-prone person: hypnosis, imagination and creativity', *Journal of Personality and Social Psychology*, Vol. 51 (1986), pp. 404–8; and Bastick, Tony, *Intuition: How We Think and Act* (Chichester: Wiley, 1982).

23. Hillman, James, *Insearch: Psychology and Religion* (Dallas, TX: Spring, 1967); and *Archetypal Psychology: A Brief Account* (Dallas: Spring, 1983).

24. Lawrence, D. H., 'Making pictures', in Ghiselin, op cit.

25. Quoted by Zervos, C., 'Conversation with Picasso', in Ghiselin, op cit.

Chapter 6

1. From Goodman, N. G. (ed), *A Benjamin Franklin Reader* (New York: Crowell, 1945). Quoted in Wilson, Timothy and Schooler, Jonathan, 'Thinking too much: introspection can reduce the quality of preferences and decisions', *Journal of Personality and Social Psychology*, Vol. 60 (1991), pp. 181–92. The experiments discussed in the first part of this chapter come from several papers by Schooler and his associates, including this one, and: Schooler, Jonathan and Engstler-Schooler, Tonya, 'Verbal overshadowing of visual memories: some things are better left unsaid', *Cognitive Psychology*, Vol. 22 (1990), pp. 36–71; Schooler, Jonathan, Ohlsson, Stellan and Brooks, Kevin, 'Thought beyond words: when language overshadows insight', *Journal of Experimental Psychology: General*, Vol. 122 (1993), pp. 166–83; and Schooler, Jonathan and Melcher, Joseph, 'The ineffability of insight', in Smith *et al* (eds), *The Creative Cognition Approach*, op cit.

2. Raiffa, H., *Decision Analysis* (Reading, MA: Addison Wesley, 1968). Quoted in Wilson and Schooler, op cit.

3. The solutions to the two 'insight' problems are:

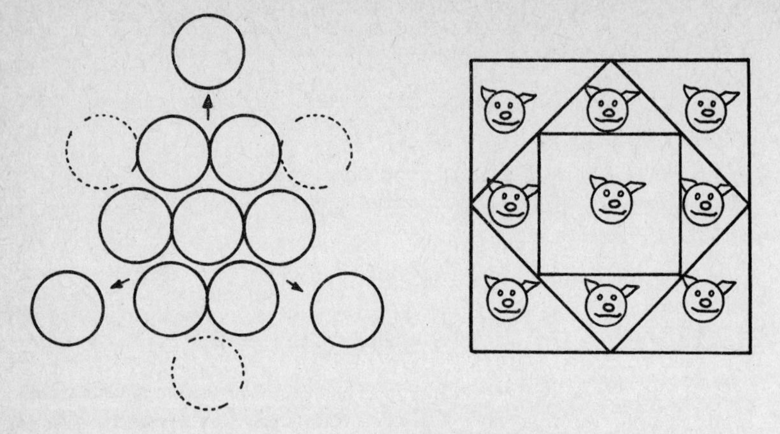

Figure 13. Solutions to coin and pig-pen problems, page 89.

The solutions to the two 'analytical' problems are:
a) The three playing cards, left to right, are:
Jack of Hearts; King of Diamonds; Queen of Spades
b) Bob is telling the truth. Alan committed the crime.

4. Things, as always, are more complicated than we would like them to be. Shifting from d-mode to responding 'impulsively' also has its risks, especially in situations that are emotionally charged. Daniel Goleman in *Emotional Intelligence* (op cit) has pointed out that a habit of impulsiveness leaves us open to being 'emotionally hi-jacked by reflexes that may have served us well in the jungle, but which can now be extremely dangerous and counter-productive.'

5. Henri Poincaré, quoted in Ghiselin, op cit.

6. Mozart, 'A letter', quoted in Ghiselin, op cit.

7. Dryden, John, 'Dedication of "The Rival-Ladies",' quoted in Ghiselin, op cit.

8. Wordsworth, William, 'Preface to Second Edition of Lyrical Ballads', quoted in Ghiselin, op cit.

9. Moore, Henry, 'Notes on sculpture', quoted in Ghiselin, op cit.

10. Gerard, R. W., 'The biological basis of the imagination', *Scientific Monthly* (June 1946).

11. Edelman, Gerald, *Neural Darwinism: The Theory of Neuronal Group Selection* (New York: Basic Books, 1992). I shall postpone any further discussion of brain mechanisms till Chapter 9.

12. Lowell, Amy, quoted in Ghiselin, op cit.

13. Housman, A. E., 'The Name and Nature of Poetry', quoted in Ghiselin, op cit.

14. Belenky, Mary Field, Clinchy, Blythe McVicker, Goldberger, Nancy Rule and Tarule, Jill Mattuck, *Women's Ways of Knowing: The Development of Self, Voice, and Mind* (New York: Basic Books, 1986).

15. Weil, Simone, *Gravity and Grace* (London: Routledge, 1972). Quoted by Belenky *et al*, op cit, p. 99.

Chapter 7

1. The existence of unconscious perception is almost universally accepted now by cognitive scientists. There are one or two die-hards, such as Douglas Holender, who wrote a long review article in 1986 attempting to find methodological flaws in every study that claimed to show unconscious perception. Norman Dixon, the doyen of the field of subliminal processing, concluded his response to Holender's paper by saying: 'the most interesting phenomenon to which Holender's paper draws attention is the extraordinary antipathy some people still have toward the idea that we might be influenced by things of which we are unaware. Would it be putting it too strongly to say it reminds one of the skepticism of "flat earth theorists" when confronted with the alarming theory that the world is round?' See Holender, D., 'Semantic activation without conscious identification in dichotic listening, parafoveal vision and visual masking: a survey and appraisal', *Behavioral and Brain Sciences*, Vol. 9 (1986), pp. 1–23; and Dixon, N.F., 'On private events and brain events', *Behavioral and Brain Sciences*, Vol. 9 (1986), pp. 29–30.

2. This study is described by Pittman, Thane, 'Perception without awareness in the stream of behavior: processes that produce and limit nonconscious biasing effects', in Bornstein, R. F. and Pittman, T. S. (eds), *Perception without Awareness: Cognitive, Clinical and Social Perspectives* (New York: Guildford Press, 1992).

3. This phenomenon is discussed in a paper by Simpson, Brian, 'The escalator effect', *The Psychologist*, Vol. 5 (1992), pp. 462–3.

4. Sidis, B., *The Psychology of Suggestion* (New York: Appleton, 1898), quoted by Merikle, P. M. and Reingold, E. M., 'Measuring unconscious perceptual processes', in Bornstein and Pittman, op cit.

5. Pierce, C. S. and Jastrow, J., 'On small differences in sensation', *Memoirs of the National Academy of Science*, Vol. 3 (1884), pp. 75–83. Quoted by Kihlstrom, J. F., Barnhardt, T. M. and Tataryn, D. J., 'Implicit perception', in Bornstein and Pittman, op cit.

6. Both the Poetzl studies and the more recent follow-ups are described

in Ionescu, M. D. and Erdelyi, M. H., 'The direct recovery of subliminal stimuli', in Bornstein and Pittman, op cit.

7. Bradshaw, John, 'Peripherally presented and unreported words may bias the perceived meaning of a centrally fixated homograph', *Journal of Experimental Psychology*, Vol. 103 (1974), pp. 1200–2.

8. Patton, C. J., 'Fear of abandonment and binge eating: a subliminal psychodynamic activation investigation', cited in Masling, Joseph, 'What does it all mean?', in Bornstein and Pittman, op cit. The fact that it may be hard consciously to accept that the subliminal perception of 'Mummy is leaving me' can have such a dramatic effect on behaviour is itself further evidence of Patton's effect.

9. Darley, J. M. and Gross, P. H., 'A hypothesis-confirming bias in labeling effects', *Journal of Personality and Social Psychology*, Vol. 44 (1983), pp. 20–33.

10. Whittlesea, B. W., Jacoby, L. L., Girard, K. A., 'Illusions of immediate memory: evidence of an attributional basis for feelings of familiarity and perceptual quality', *Journal of Memory and Language*, Vol. 29 (1990), pp. 716–32.

11. Schacter, Daniel (ed.), *Memory Distortions: How Minds, Brains and Societies Reconstruct the Past* (Cambridge, MA: Harvard University Press, 1995).

12. I shall say more about the way in which we can protect ourselves from our own unconscious assumptions in the discussion of 'mindfulness' in Chapter 11.

13. See Nisbett, R. and Wilson, T., 'Telling more than we know: verbal reports on mental processes', *Psychological Review*, Vol. 84 (1977), pp. 231–59.

14. Latane, B. and Darley, J. M., *The Unresponsive Bystander: Why Doesn't He Help?* (New York: Appleton-Century-Crofts, 1970).

15. Fitzgerald, F. Scott, *Tender is the Night* (New York: Scribner, 1934).

16. The case of Flournoy and Helen Smith is discussed by Ellenberger, Henri, *The Discovery of the Unconscious* (New York: Basic Books, 1970), p. 316.

Chapter 8

1. Masling, Joseph M., 'What does it all mean?', in Bornstein and Pittman, op cit.

2. Bruner, Jerome and Postman, Leo, 'Emotional selectivity in perception and reaction', *Journal of Personality*, Vol. 16 (1947), pp. 69–77.

3. It is surprising how frequently well-educated adults in our society fear that any kind of psychological trick or test is liable to expose something unwelcome about their mental powers. Television quiz shows, for example, both reflect and promote the ridiculous assumption that rapid retrieval of trivia is a valid index of 'intelligence' – though schools, with somewhat more pretension, may fall into the same trap.

4. A similar interpretation of the Zajonc studies has been offered by Reber, op cit.

5. Quoted by Reber, op cit, p. 18.

6. These studies of 'implicit memory', as it is called, are comprehensively reviewed by Schacter, Daniel, 'Implicit memory: history and current status', *Journal of Experimental Psychology: Learning, Memory and Cognition*, Vol. 13 (1987), pp. 501–18.

7. Marcel, Tony, 'Slippage in the unity of consciousness', in CIBA Symposium 174, *Experimental and Theoretical Studies of Consciousness* (Chichester: Wiley, 1993).

8. Cumming, Geoff, 'Visual perception and metacontrast at rapid input rates', DPhil thesis, University of Oxford (1971).

9. Marcel, op cit.

10. The best discussions of 'functional blindness' are still those provided by P. Janet in his classic text *The Major Symptoms of Hysteria* (New York: Macmillan, 1907). A marvellous fictionalised description is provided by William Wharton in *Last Lovers* (London: Granta, 1991).

11. See Wall, Patrick, in CIBA Symposium 174, op cit.

12. Sutcliffe, J. P., '"Credulous" and "skeptical" views of hypnotic phenomena: experiments in esthesia, hallucination and delusion', *Journal of Abnormal and Social Psychology*, Vol. 62 (1961), pp. 189–200.

13. Langer, E., Dillon, M., Kurtz, R. and Katz, M., 'Believing is seeing', unpublished paper, Harvard University, referred to in Langer, Ellen, *Mindfulness: Choice and Control in Everyday Life* (London: Harvill, 1991).

14. For a review of blindsight studies, see Weiskrantz, Lawrence, *Blindsight: A Case Study and Its Implications* (Oxford: Clarendon, 1986).

15. Humphrey, Nicholas, comments in discussion, in CIBA Symposium 174, op cit, p. 161.

16. I once asked Tony Marcel whether a thirsty blindsight patient would spontaneously reach for a glass of water that was within the blind field. In practice, he pointed out, this would be an almost impossible test to carry out, as none of these patients is 'blind' in all areas of the visual field, so when their eyes are free to move about, as they normally are, any significant

objects in their world would rapidly be picked up through the areas of normal sight. But (for what it's worth) his strong intuition, having worked with such patients for some time, is that they would not.

17. Freud, Sigmund, 'Recommendations to physicians practising psychoanalysis', in Strachey, J. (ed. and trans.), *The Standard Edition of the Complete Psychological Works of Sigmund Freud*, Vol. 12 (London: Hogarth Press, 1958/1912).

18. Granger, G. W., 'Night vision and psychiatric disorders', *Journal of Mental Science*, Vol. 103 (1957), pp. 48–79.

19. Bahrick, H. P., Fitts, P. M. and Rankin, R. E., 'Effect of incentives upon reactions to peripheral stimuli', *Journal of Experimental Psychology*, Vol. 44 (1952), pp. 400–6.

20. Bursill, A. E., 'The restriction of peripheral vision during exposure to hot and humid conditions', *Quarterly Journal of Experimental Psychology*, Vol. 10 (1958), pp. 113–29.

21. Bruner, J. S., Matter, J. and Papanek, M. L., op cit.

Chapter 9

1. Dickinson, Emily, 'The Brain', in *Complete Poems* (Boston: Little, Brown, 1960), reprinted in Mitchell, S. (ed), *The Enlightened Heart* (New York: Harper & Row, 1989).

2. We now know that the three 'systems' are so tightly integrated with each other that it is more accurate to see them as three aspects of what is in effect a single system. Indeed, if we are to appreciate the physical underpinnings of the slow ways of knowing we need to reinstall the brain in its bodily context, and we shall do so in Chapter 10. But it makes sense to start with the brain on its own.

3. So far the cells that have shown LTP, in an area of the midbrain called the hippocampus, do tend to revert to their original recalcitrant state eventually, so they themselves cannot be responsible for the memories of a lifetime. It will not be long before some similar but even more permanent mechanism is found that will weld together cells in the cortex. But for the moment this remains just beyond the leading edge of what it is technically possible to investigate.

4. Hebb, D. O., *The Organization of Behavior* (New York: McGraw Hill, 1949).

5. See for example Minsky, Marvin, *The Society of Mind* (London: Picador, 1988).

6. This evidence is reviewed in Greenfield, Susan, *Journey to the Centers of*

the Mind (Oxford: Freeman, 1955). My image of brain organisation, a preliminary version of which was first published in my *Cognitive Psychology: New Directions* (London: Routledge, 1980) is in many respects similar to Susan Greenfield's, a coincidence that may be not unrelated to the fact that we were graduate students together in Oxford in the early 1970s. The major differences are that my model tries to find a place for language; and our views on the role of arousal are somewhat divergent.

Chapter 10

1. This evidence is reviewed in Martindale, Colin, 'Creativity and connectionism', in Smith, Ward and Finke, op cit.

2. This illustration is a development of one used by Edward de Bono in *The Mechanism of Mind* (Harmondsworth: Penguin, 1971).

3. Luria, A. R., *The Mind of a Mnemonist* (Harmondsworth: Penguin, 1975).

4. There is some direct evidence for this obvious assumption: see Grossberg (1980), cited in Martindale, op cit; Kahneman, D., *Attention and Effort* (Englewood Cliffs, NJ: Prentice Hall, 1973); Baddeley, A. D. and Weiskrantz, L. (eds), *Attention: Selection, Awareness and Control* (Oxford: Clarendon, 1993).

5. The influences of culture and experience cannot in practice be separated in the neat way that this picture might imply. Much if not all of a child's direct experience with the words is both mediated by 'agents' of the culture, and saturated with cultural assumptions. Parents, older siblings and teachers are continually guiding a young person's attention, implicitly instructing her as to what is worth attention, and what significance these selected experiences are to be assigned. (Children are quick to pick up value judgements – such as phobias, for example – from observing the reactions of their seniors.) And even if no agent is physically present, the child's world is full of objects and experiences that embody the values and assumptions of the culture: toys, games, artifacts and rituals of all kinds. Even the buildings which she inhabits, and the physical landscape through which she moves, are repositories of cultural meaning.

The two-plane model which I am using here is a simple version of the kind of 'hybrid' model that many neural network theorists are currently exploring. See for example Churchland, P. S. and Sejnowski, T. J., *The Computational Brain* (Cambridge, MA: Bradford/MIT Press, 1992).

6. I am drawing on some of Gelernter's ideas in the paragraphs that follow. See Gelernter, David, *The Muse in the Machine: Computers and Creative Thought* (London: Fourth Estate, 1994).

7. Dennett, Daniel, *Consciousness Explained* (London: Viking, 1992).

8. Young, A. W. and De Haan, E. H., 'Face recognition and awareness after brain injury', in Milner, A. D. and Rugg, M. D. (eds), *The Neuropsychology of Consciousness* (London: Academic Press, 1992).

9. Research quoted by Greenfield, op cit.

10. Libet, Benjamin, 'The neural time factor in conscious and unconscious events', in CIBA Symposium 174, op cit.

11. Experiment by Jensen (1979), quoted by Libet, op cit, p. 126.

12. I have elaborated this argument in my *Noises from the Darkroom: the Science and Mystery of the Mind* (London: HarperCollins, 1994).

13. For an elaboration of this argument, see my article 'Structure, strategy and self in the fabrication of conscious experience', *Journal of Consciousness Studies*, Vol. 3 (1996), pp. 98–111.

14. Kihlstrom, John, 'The psychological unconscious and the self', in CIBA Symposium 174, op cit, p. 152.

15. Libet, op cit.

16. I have argued this point of view in more detail in my *Noises from the Darkroom: The Science and Mystery of the Mind*, op cit; as also has Velmans, Max, 'Is human information processing conscious?', *Behavioral and Brain Sciences*, Vol. 14 (1992), pp. 651–726; and Mandler, George, *Mind and Emotion* (New York: Wiley, 1975). The argument is very similar to Keith Oatley's, in *Best Laid Schemes* (Cambridge: CUP, 1992).

17. Churchland, Patricia, *Neurophilosophy* (Cambridge, MA: MIT Press, 1986).

Chapter 11

1. Ginzburg, Carlo, *Myths, Emblems, Clues* (London: Hutchinson Radius, 1990). I am grateful to Alan Bleakley for putting Ginzburg's work my way.

2. Cited in Ginzburg, op cit, p. 211.

3. Conan Doyle, Sir Arthur, 'The Cardboard Box', first published in the *Strand* magazine, Vol. 5 (1893), pp. 61–73. Quoted by Ginzburg, op cit.

4. Freud, Sigmund, 'The Moses of Michelangelo' (1914), in *Collected Papers* (New York: Hogarth Press, 1959). Quoted by Ginzburg, op cit.

5. Reiser, Stanley, *Medicine and the Reign of Technology* (Cambridge: CUP, 1978). Quoted in Postman, op cit.

6. From Seltzer, Richard, *Mortal Lessons* (New York: Simon and Schuster, 1974). Reprinted in Feldman, Christina and Kornfield, Jack (eds), *Stories of the Spirit, Stories of the Heart* (San Francisco: HarperCollins, 1991).

7. This research, as well as details of the focusing process, are described in Gendlin, Eugene, *Focusing* (New York: Bantam, 1981).

8. I can vouch both for the effectiveness of focusing, and for its subtle, slippery quality, as I have taken two training courses in it. Some people find it easier and quicker to grasp than others, and it needs coaching, feedback and modelling, as well as direct tuition, if one is to get the hang of it. Learning to 'focus' is of the same order of difficulty as any other form of delicate perceptual learning – wine-tasting, reading X-rays or animal tracks, and so on.

9. Gendlin, op cit.

10. Suzuki, D. T., *Zen and Japanese Culture* (Princeton, NJ: Princeton University Press, 1959), pp. 104–5, 109, 157.

11. Dodds, E. R., *The Greeks and the Irrational* (Berkeley, CA: University of California Press, 1951). See also Onians, R. B., *The Origins of European Thought* (Cambridge: CUP, 1951).

12. This feeling, referred to as *yugen*, is much prized by the Zen-inspired painters and poets of Japan. The poet Seami says *yugen* is 'To watch the sun sink behind a flower-clad hill, to wander on and on in a huge forest with no thought of return, to stand on the shore and gaze after a boat that goes hid by far-off islands, to ponder on the journey of wild geese seen and lost among the clouds.' To which Alan Watts, in *Nature, Man and Woman* (London: Thames and Hudson, 1958), adds: 'But there is a kind of brash mental healthiness ever ready to rush in and clean up the mystery, to find out just precisely where the wild geese have gone . . . and that sees the true face of a landscape only in the harsh light of the noonday sun. It is just this attitude which every traditional culture finds utterly insufferable in Western man, not just because it is tactless and unrefined, but because it is blind. It cannot tell the difference between the surface and the depth. It seeks the depth by cutting into the surface. But the depth is known only when it reveals itself, and ever withdraws from the probing mind.'

13. Cassirer, Ernst, *Language and Myth* (New York: Harper, 1946).

14. Quoted by Scott, Nathan, *Negative Capability: Studies in the New Literature and the Religious Situation* (New Haven, CT: Yale University Press, 1969).

15. Gardner, Howard and Winner, Ellen, 'The development of metaphoric competence: implications for humanistic disciplines', in Sacks, S. (ed.), *On Metaphor* (Chicago: University of Chicago Press, 1979).

16. Dimnet, Ernest, quoted in de la Mare, Walter, *Behold this Dreamer!* (London: Faber & Faber, 1939), p. 647.

17. Maritain, Jacques, *Creative Intuition in Art and Poetry* (London: Harvill, 1953).

18. Quoted by Scott, op cit.

19. Eliot, T. S., *Four Quartets* (London: Faber & Faber, 1959).

20. From John Anderson's introduction to Heidegger's *Discourse on Thinking* (New York: Harper and Row, 1966)

21. Rilke, Rainer Maria, *Letters to a Young Poet*, translated and introduced by R. Snell (London: Sidgwick, 1945).

22. Whalley, George, 'Teaching poetry', in Abbs, Peter (ed), *The Symbolic Order* (London: Falmer Press, 1989), p. 227.

23. Housman, A. E., quoted in Ghiselin, op cit.

24. Croce, Benedetto, *Aesthetic*, translated by Ainslie Douglas (New York: Noonday/Farrar, Straus, 1972).

25. MacNeice, Louis, 'Snow', reprinted in Allott, Kenneth (ed), *The Penguin Book of Contemporary Verse* (Harmondsworth: Penguin, 1962). Interestingly, in the context of the present discussion, Allott comments of MacNeice: 'He is too eager and impatient to accept his subject quietly and try to understand it. He grabs it, pats it into various shapes, and varnishes any cracks in the quality of his perception with his prestidigitatory skill with words and images.'

26. Borges, Jorge Luis, *Labyrinths* (Harmondsworth: Penguin, 1970).

27. Sacks, Oliver, 'Rebecca', in *The Man who Mistook his Wife for a Hat* (London: Duckworth, 1985) pp. 169–77.

28. Kanizsa, G., *Organisation of Vision: Essays in Gestalt Psychology* (New York: Praeger, 1979).

29. This and several of the other illustrations in this chapter are taken from Langer, Ellen, *Mindfulness*, op cit.

30. Holmes, D. and Houston, B. K., 'Effectiveness of situation redefinition and affective isolation in coping with stress', *Journal of Personality and Social Psychology*, Vol. 29 (1979), pp. 212–18.

31. Teasdale, John, Segal, Zindel and Williams, Mark, 'How does cognitive therapy prevent depressive relapse and who should attentional control (mindfulness) training help?', *Behavioral Research and Therapy*, Vol. 33 (1995), pp. 25–39.

32. Teasdale *et al*, op cit.

33. Goleman, Daniel, *Emotional Intelligence*, op cit.

Chapter 12

1. This incident occurs in a film made about Summerhill in the 1970s by the Canadian Film Board.

2. This story is told in Watzlawick, Paul, Weakland, John, and Fisch, Richard, *Change: Principles of Problem Formation and Problem Resolution*, op cit.

3. Labouvie-Vief, Gisela, 'Wisdom as integrated thought: historical and developmental perspectives', in Sternberg, R. J. (ed), *Wisdom: its Nature, Origins and Development* (Cambridge: CUP, 1990).

4. Kekes, J., quoted by Kitchener, Karen and Brenner, Helene, 'Wisdom and reflective judgement', in Sternberg, op cit.

5. Robin Skynner discussed this in a seminar with Fritjof Capra at Schumacher College, Devon, in June 1992.

6. Rogers, Carl, *A Way of Being* (New York: Houghton Mifflin, 1981).

7. Kierkegaard, Soren, quoted by Pascual-Leone, Juan, 'An essay on wisdom: toward organismic processes that make it possible', in Sternberg, op cit.

8. Sternberg, Robert J., 'Implicit theories of intelligence, creativity and wisdom', *Journal of Personality and Social Psychology*, Vol. 49 (1985), pp. 607–27.

9. Kegan, Robert, *In over our Heads: the Mental Demands of Modern Life* (Cambridge, MA: Harvard University Press, 1994).

10. Meacham, John, 'The loss of wisdom', in Sternberg, *Wisdom*, op cit.

11. In this and the other quotations in this section, the emphases have been added to the originals.

12. Quotation and details of Tauler's life taken from Moss, Donald M., 'Transformation of self and world in Johannes Tauler's mysticism', in Valle, R. S. and von Eckartsberg, R. (eds), *The Metaphors of Consciousness* (New York: Plenum Press, 1981).

13. Whyte, Lancelot Law, *The Unconscious before Freud* (London: Julian Friedmann, 1978), p. 10.

14. Excerpts from *Free and Easy: A Spontaneous Vajra Song* by Lama Gendun Rinpoche.

15. Suzuki, Shunryu, *Zen Mind, Beginner's Mind* (New York: Wetherhill, 1970).

16. Sahn, Seung, *Dropping Ashes on the Buddha*, S. Mitchell (ed) (New York: Grove Press, 1976).

17. Quotations drawn from Suzuki, D. T., *The Zen Doctrine of No Mind* (London: Rider, 1969); and Yampolsky, Philip, *The Platform Sutra of the Sixth Patriarch* (New York: Columbia University Press, 1967).

Chapter 13

1. Jaynes, Julian, *The Origin of Consciousness in the Breakdown of the Bicameral Mind* (Boston: Houghton Mifflin, 1976).

2. Dodds, op cit.

3. These quotations are drawn from Whyte, op cit.

4. Ibid, pp. 41–2.

5. Postman, op cit.

6. Ibid. p. 111.

7. Ibid. pp. 118–19.

8. From Heidegger, Martin, *Discourse on Thinking*, op cit.

9. Description of the GMAT in the 1996–7 GMAT Bulletin, published by the Graduate Management Admission Council, Princeton, NJ.

10. See the American Psychological Association review of 'Intelligence: knowns and unknowns', chaired by Ulric Neisser, published in *American Psychologist*, Vol. 51 (1996), pp. 77–101.

11. Ceci and Liker, op cit.

12. Peters, Tom, *The Pursuit of Wow! Every Person's Guide to Topsy-Turvy Times* (New York: Vintage, 1994).

13. Peters, Tom, 'Too wired for daydreaming', *Independent on Sunday*, 13 February 1994.

14. Peters, *The Pursuit of Wow!*, op cit.

15. Rowan, Roy, *The Intuitive Manager* (Boston: Little, Brown, 1986).

16. De Bono, Edward, *De Bono's Thinking Course* (London: BBC, 1985).

17. Rowan, op cit.

18. Mintzberg, Henry, *The Rise and Fall of Strategic Planning* (New York: The Free Press, 1994).

19. Quinn, Brian, quoted by Mintzberg, op cit.

20. See de Bono, op cit.

21. West, Michael, Fletcher, Clive and Toplis, John, *Fostering Innovation: A Psychological Perspective* (Leicester: British Psychological Society, 1994).

22. For a summary of Dweck's work, see Chiu, C., Hong, Y. and Dweck, C. S., 'Toward an integrative model of personality and intelligence: a general framework and some preliminary steps', in Sternberg, R. J. and Ruzgis, P. (eds), *Personality and Intelligence* (Cambridge: CUP, 1994).

23. For further examples and discussions of science education, see Claxton, Guy, *Educating the Inquiring Mind: The Challenge for School Science* (Hemel Hempstead: Harvester/Wheatsheaf, 1991); Claxton, Guy, 'Science of the times: a 2020 vision of education', in Levinson, R. and Thomas, J. (eds), *Science Today: Problem or Crisis*? (London: Routledge, 1996); Cosgrove, Mark, 'A study of science-in-the-making as students generate an analogy for electricity', *International Journal of Science Education*, Vol. 17 (1995), pp. 295–310; and Osborne, Roger and Freyberg, Peter (eds), *Learning in Science* (Auckland and London: Heinemann, 1985).

24. Archbishop William Temple, quoted in Watts, Alan, *In my Own Way* (New York: Vintage, 1973).

25. See for example Gallwey, Timothy, *The Inner Game of Tennis* (London: Cape, 1975); Clark, Frances Vaughan, 'Exploring intuition: prospects and possibilities', *Journal of Transpersonal Psychology*, Vol. 3 (1973), pp. 156–69.

26. Langer, E., Hatem, M., Joss, J. and Howell, M., 'Conditional teaching and mindful learning: the role of uncertainty in education', *Creativity Research Journal*, Vol. 2 (1989), pp. 139–50.

27. See Nisbet, J. and Shucksmith, J., *Learning Strategies* (London: Routledge, 1986).

28. Whyte, op cit.

Index

Numbers in italics refer to diagrams.